The Good Pope

THE GOOD POPE

The Making of a Saint
and the Remaking of the Church—
The Story of John XXIII and Vatican II

GREG TOBIN

HarperOne
An Imprint of HarperCollinsPublishers

HarperOne

Frontispiece credit: Gamma-Keystone/Getty Images.

FIRST HARPERCOLLINS PAPERBACK EDITION PUBLISHED IN 2013

Library of Congress Cataloging-in-Publication Data

Tobin, Greg.
    The good Pope : the making of a saint and the remaking of the church —
the story of John XXIII and Vatican II / by Greg Tobin.
        p. cm.
    ISBN 978-0-06-208941-0
    1. John XXIII, Pope, 1881-1963. 2. Popes—Biography. 3. Vatican Council
(2nd : 1962-1965) 4. Catholic Church—History—20th century. I. Title.
BX1378.2.T58   2012
282.092—dc23
    [B]   2012017653

13 14 15 16 17   RRD(H)   10 9 8 7 6 5 4 3 2 1

To all women and men of good will,
in the spirit and memory of the Good Pope

Contents

Contents

PART III: FATHER OF THE COUNCIL

The Making of a Saint and the Remaking of a Church

They called him *Il Buono Papa,* "the Good Pope." During Pope John XXIII's lifetime—and especially in the immediate aftermath of his death from stomach cancer on June 3, 1963—Italian Catholics and Socialists alike; journalists and diplomats; Roman Catholics, Protestants, non-Christians, and nonbelievers across the globe; men and women of every race, class, and nation called him "good" and mourned his passing. His doughy peasant's face, his dark eyes, ample jowls, beaked nose, and big-lobed ears were known throughout the world. He was a megawatt celebrity in the age of such secular saints as Elizabeth Taylor and Richard Burton, John Fitzgerald Kennedy and Jackie Kennedy, Fidel Castro and Nikita Khrushchev. But there was nothing glamorous about Angelo Giuseppe Roncalli. Even in his papal vestments he carried himself like the son of Italian farmers, themselves the sons and grandsons of countless generations of farmers.

For those who admired him during his lifetime for his teachings on peace and his commitment to open his ancient Church to the modern world—to allow air and light in and let the profound message of the gospel shine out—he stood as a unique character radiating an aura of humility, humor, and sanctity. His genius lay in the willingness to introduce the concept of *aggiornamento,* or updating, so that his beloved Church might benefit from an infusion of "fresh air" through newly opened "windows."

Now, a half-century after his death, he is almost certain to be canonized, officially recognized by his Church, as a saint. In 2000, one of his successors, Pope John Paul II, named him "Blessed" in the penultimate step to sainthood, ensuring that this remarkable figure would be remembered—and revered—by the generations.

But why, beyond his popularity and unusual charisma—his spiritual celebrity, if you will—should he be singled out for such honor and glory, which he would undoubtedly eschew, perhaps angrily if he were here among us to speak for himself?

Some figures in Christianity and the history of the Catholic Church stand out in bold relief for their holiness in life, their bold or unusual personalities, and the imprint they left behind as servants of the faith: Thomas Aquinas, Edith Stein, Pope Gregory the Great, Mother Teresa of Calcutta are models of sanctity and service who have been canonized by the Catholic Church. Many such saints left behind writings meant to instruct and inspire others—think of Augustine of Hippo and Thérèse of Liseux as two more examples. All exhibited faith, hope, and charity in their daily living. Perhaps not all day, every day, for saints are imperfect human beings, hardly angels or gods. Saints and sainthood are not necessarily to be considered rarities; rather every Christian is called to this state of sanctity. Holiness is within one's grasp at any hour of the day and all around us, for each of us, whether we choose to acknowledge and tap into it or not.

Still, how could such a man as Roncalli, known to the world as John XXIII, and who in many ways was a throwback to previous centuries, shine so brightly in the modern world? After all, he had come of age and labored mightily in a Roman Catholic Church that appeared to many to be tired, stale, and defensive in its stance against perceived enemies. How did this pious peasant priest engage the protean, secular, fast-moving contemporary world so distinctly and effectively? And how was he able to instill his Church with a newfound confidence to become the shepherd for the whole world and dramatically open the doors of the Church to meaningful, necessary, and faithful change?

This biography is my attempt to answer these questions. And I feel some urgency in finding these answers, since it could be argued that many of the crises and tensions within the Catholic Church today result from later leaders moving away from what John modeled and accomplished. The fault lines that exist in the contemporary Church are different but similar to previous divides that have scored the institution in centuries past—including scandals and theological debates—but there is no singular figure or movement on the scene today to compare to Roncalli.

The eternal question that has concerned the Church since the first Pentecost in Jerusalem is, Can the institution founded in the apostolic age survive another generation in the face of worldly forces of opposition and human failures and corruption within its own ranks?

Another way to ask these questions is to ask their opposite: What would have happened if an ultratraditionalist or a less pastoral-minded cardinal had been elected as pope? The answer seems obvious: the entire liberalizing movement of the 1960s and the Church's response to it might have wrenched Catholics too much in one direction or another and not presented as open a face to a rapidly changing world. As it turns out, that face conveyed both a bedrock, conventional religious piety as well as a remarkable and hitherto unfamiliar tolerance for theological and political currents that could perhaps have broken a more rigid, pharaoh-like leader. John, stout as he was, sat lightly upon the throne, beneath the cumbersome triple tiara—more so than perhaps any of his predecessors in history—at exactly the moment when such an attitude would be critical for the very survival and creative adaptation of the Catholic idea in a brave new world of burgeoning technology, mass communication, and deadly competition among cultures virtually everywhere on Earth.

It is fair to say that the Church and the world would have been much different if John had not been elected in 1958. Arguably, he slowed the escalation of the nuclear arms race, at least temporarily, at a crucial juncture in the Cold War through his calming presence during the Cuban missile crisis, his teaching in the encyclical *Pacem in terris* (Peace on Earth), and his deft, personal touch in diplomacy, perfected over a quarter century through some of the thorniest assignments imaginable for an apostolic nuncio in Europe in the decades leading up to and during World War II.

Setting the tone for his successors, he moved the Church in a new direction in its relationship to Jews, which bore fruit in the Vatican Council's declaration *Nostra aetate* (In Our Age), and to non-Catholic Christians that ultimately resulted in the decree on ecumenism *Unitatis redintegratio* (The Restoration of Unity) and a whole new attitude on the part of clergy, the hierarchy, and the laity.

This pope still matters because he stood with his feet planted firmly in the swiftly flowing river of history and, like the legendary Saint Christopher, helped his people move safely from one bank to the other without being swept away by the raging currents beneath. Thus he "saved" the Church he loved so much, preserving its core doctrines intact, through force of will and personal diplomacy as manifested in a humble, indeed earthy spirituality that contradicted most expectations by his peers (he would call them, not without irony, his "betters" in the Church hierarchy).

He was thus able to move the immovable and open up possibilities for reform—or even a discussion of reform—that his immediate predecessors and contemporaries would not even consider. Perhaps only Pius X (though no liberal he, and the last pope to date to be canonized) could match John for his reforming pastoral zeal.

Furthermore, John did not allow the pomp of papal ceremony or the inertia of a legendary millennia-old bureaucracy to divert him from his agenda. He knew he had but a limited time to accomplish his ambitious program, symbolized and given substance by the Second Ecumenical Council of the Vatican. Indeed, when he was advised, early in the planning phase in 1960, that it was unlikely the council could be convened before 1964, he promptly decreed that it would open in 1962. It did.

John would often surprise his peers and superiors. Yet they continued to underestimate him throughout his life. Even today, five decades after his death, some within the Church deride him as "popular" or "soft"—though the vast majority of Catholics and a large number of Anglicans, Protestants, Jews, and other non-Christians think highly of John XXIII and his legacy of tolerance and openness.

Most Catholics, except a diehard few who view Vatican II reforms as a heretical schism within the True Church of Christ (and there are those who do—a quick tour of Internet sites will confirm this), expect that he will be canonized soon—very soon.

This book, then, represents an attempt to understand *why* the phenomenon of Pope John XXIII and his reputation as *Il Buono Papa*, the Good Pope, remains so durable and so inspirational for so many and why he deserves to be known by new generations of Catholics in search of a more open and ecumenical Church. And, I posit, why he will long outlast the critics and opponents of his council still among us, and those yet to come.

THE GOOD POPE

PART I

Priest and Protector

CHAPTER ONE

Pastor et Nauta,
Shepherd and Navigator

The eight other modern popes—Leo XIII (1878–1903), Saint Pius X (1903–14), Benedict XV (1914–22), Pius XI (1922–39), Pius XII (1939–58), Paul VI (1963–78), John Paul II (1978–2005), and Benedict XVI (2005–present)—exerted their distinctive influence on the Church they had inherited in their time. (John Paul I occupied the papacy for only thirty-three days in 1978.) Yet for the most part their concerns were internal to the Church, which had been buffeted, battered, and split during nineteen centuries. So many of their predecessors down through the ages had built up defensive walls and sent out armies to conquer rather than open doors and persuade souls to come to the gentle Christ. These modern popes, indeed, faced the myriad threats to the faith with no temporal power at their disposal and diminished spiritual authority, though they tried valiantly (albeit not always successfully) to regain the moral ground on which prior successors of Saint Peter had stood.

Pope John XXIII came to the throne in a world that had survived Nazism and the Holocaust—if barely—and a devastating world war prior to that. Besides fascism and Nazism, he confronted communism and the Cold War, the atomic arms race, and civil strife in many corners of the earth. Science and technology were taking great leaps forward in the West, and the population of the planet was exploding, especially in the undeveloped nations or "Third World." He was quite well attuned to and prepared for this world. Throughout most of his forty-year career, he worked in the world, far removed from the oftentimes claustrophobic papal court of Rome.

His Church had survived two world wars, and many within its hierarchical leadership felt that it should retrench in the face of "enemies" such as communism and modernism—in effect refighting the battles of previous centuries but without any new weapons in its armory.

His status as a non-Roman, from outside the inner circles of power, made him not only a long shot for election to the papacy but an unlikely candidate for revolutionary. He was seventy-seven when he was elected the Roman pontiff. As a diplomat in Bulgaria, Greece, and Turkey in the years leading up to and during World War II, he had not overly impressed his teachers or army commanders, nor his masters in the Vatican. "Unlikely" may as well have been tattooed on his forehead.

Yet a lifetime's varied experience had taken hold of Angelo Giuseppe Roncalli: his life had been a rich synthesis of places, people, and circumstances. He had grown up among the farmer-peasants of Bergamo; served with down-to-earth army men; worked with fund-raisers for a curial congregation within the Vatican; spoken with beleaguered Orthodox Christians in Bulgaria; and reached out to Muslims in Turkey, collaborationists in France, and communists there and in Italy. He ministered respectfully to priest-workers, irrespective of whether they were in or out of favor in Rome.

Pragmatic and friendly, he kept the flame of his spiritual life lit always through prayer and quiet good works—and with an intelligent, ironic humor. Intuitive, he was schooled in sacred theology and ecclesial history but was never overtly intellectual; rather he navigated human relationships within and outside the Church with skills and instincts honed over decades in the Vatican Foreign Service.

The startling ascent of a convivial backwater diplomat from rural Italy to the supreme and most powerful office in Christianity in 1958, unlikely as it was, pales somewhat in historical significance to what followed immediately upon the coronation of Roncalli as John XXIII. When he called for a new Vatican Council—a gathering of all the bishops in one place and the first since 1870—he shattered all previous expectations that his would be a transitional or uneventful pontificate.

As he lay dying on the last day of May 1963, Roncalli received his Roman cardinals. He told each of them he was "on the point of leaving." It was as if he were setting out on a journey and would not see his friends and associates for a while. Having fulfilled his role as *pastor et nauta,* shepherd and navigator, a fisher of men, and having set the "Barque of Peter" (an old-fashioned term used to describe the Church and the papacy, in reference to the apostle's profession as a fisherman) out onto the sea of the world, he could now himself set forth on the sea of eternity, almost alone. He called for the viaticum (the last Holy Communion means Christ "with you on the way"—*a via tecum*), and soon he was off on his final voyage beyond the horizon of eternity.

But not before he told many he was offering his life "for a good outcome of the ecumenical council and for peace among men," which was, in effect, his dying wish. To the last he was quietly insistent on his basic vision. Intuitive, not always well expressed, it came across nevertheless in the Second Vatican Council and in the hopeful love for all men of good will that was expressed in *Pacem in terris.* With no ambiguity now at the end, with the lucidity of a dying man, he repeated and repeated his wish, "that the great work will be crowned with success."

Pope John XXIII was a gentle revolutionary. Far from being the caretaker that the Church expected, John created an atmosphere in which, said Jesuit theologian John Courtney Murray, "a lot of things came unstuck—old patterns of thought, behavior, feeling. They were not challenged or refuted, but just sort of dropped."

The monolithic Church of the Middle Ages would not— perhaps could not—resist the historic movement, encouraged by this pope, toward dynamism and diversity. However painfully, Mother Church *would* change. And she would then face the inevitable internal reaction and retrenchment that follows any cultural or spiritual revolution.

Where did he come from? His birth, youth, and early priesthood— his first twenty-five years—are fairly well documented, considering his unremarkable progress through life in those years. He was a child not only of the Roncalli clan of Sotto il Monte in the region of Bergamo in northern Italy, but a son of the Church, literally, from the day of his birth. There seems to be no question in anyone's mind that he was always destined for a vocation as a priest, as natural a fact of his life as breathing.

Angelo Giuseppe Roncalli came from the soil of the mountains, from the heart of a traditionally pious and typically boisterous family of twelve siblings (three of whom died very young) and numerous close relatives. He grew up under the direct and unending influence of the Roman Catholic Church of Pope Leo XIII, one of the most able and progressive pontiffs of the past two centuries. Just as young Roncalli reached his majority and shortly before his ordination as a priest, Pius X was elected to the papacy and for a dozen years stood as an antimodernist bulwark, to be succeeded by the scholarly Benedict XV on the eve of the Great War.

Young Roncalli experienced more continuity than change in his holy, enduring Church, even though the theological winds and emerging modern values blew up against the fortress that had been buttressed by the sixteenth-century Council of Trent and Pius IX's inwardly focused Council of the Vatican in 1869–70, which defined papal infallibility at that time and for all time. Piety trumped inquiry in the Church in which Roncalli came of age, but he felt encouraged to follow the path of historical research and reflection that would feed his capacious mind and fill his big heart—over time—with insights into the human and organizational aspects of his beloved institution, Holy Mother Church.

It must be remembered, however, before hoping to understand John, that doubts and speculations seemingly never entered the picture for the seminarian or the neophyte priest. There was no room in a life devoted to faith—his own, his family's, and that of those to whom he sought to minister and for whom he prayed—to entertain such worldly distractions or intellectual game playing. Life was too precious, and too short for most, and salvation was its ultimate end for the faithful Christian. Period.

And where did he go? From the parochial world of his mountain village, and even of Bergamo, a small but storied imperial city northeast of Milan with famous traditions of music, sanctity, and warfare, he would move to Rome, the living heart of the Church, then to ever-widening roles on the European diplomatic stage before emerging, late in his life, as the "Pope of the World." He also experienced World War I firsthand, in the trenches, as a young noncommissioned officer and chaplain in the Italian Army. His was a remarkable pilgrim's progress that began inauspiciously, to say the least.

He progressed, it might be said, from family to family, from his own flesh and blood to the fraternity of seminary and priesthood, to the distant "family" of the faithful in the Balkans and in cosmopolitan France, to the people of Venice for whom he was patriarch and *pater familias*. Always he tried to remain in contact with the Roncallis of Sotto il Monte, including his parents, who both died in the 1930s (when Angelo was in his fifties), siblings, and nieces and nephews. Finally, as *il Papa* and Holy Father, his family extended across the globe, to mountain villages in Africa and Asia and cities in the Americas that he could never hope to visit in person as his successors one day would.

Who and what formed his character? As a boy, it seems he was touched with holiness (a term that needs to be explained and explored in depth in the context of this man and his time) from a source outside himself. He was a good boy and a good man, by all accounts. How he became the Good Pope, then, is the story that cries out to be told to a new generation.

CHAPTER TWO

Family, Youth, and Seminary (1881–1904)

In the year of 1881, 25 November, I, Francesco Rebuzzini, the priest of this church of San Giovanni Battista of Sotto il Monte, baptized the infant born today of the lawfully married couple Giovanni Battista Roncalli and Marianna Mazzola, from Brusico in this parish. The infant was given the names Giovanni Giuseppe.

—PARISH REGISTER

This simple entry in the parish register of a northern Italian hill town is the first public record of the future Pope John XXIII, and the error it contains—the priest recorded the infant's name as Giovanni Giuseppe when it was actually Angelo Giuseppe—highlights a number of the circumstances surrounding the child's birth.

Tucked away in the foothills of the Alps, just north of the Lombard plains, the village of Sotto il Monte—whose name means "Under the Mountain"—was a nondescript, out-of-the-way place. A cluster of gray stone buildings, it was home to about 1,200 people, most of whom made their living from the land. The weather in Sotto il Monte was often inclement—hot summers alternating with wet winters, when fierce winds swept down from the Alps, bringing rain that made roads nearly impassable with mud.

Angelo Roncalli was born on one such day, when the *tramontano*—the northerly mountain gale that can reach speeds of 50 miles per hour—buffeted the town. He came into the world around ten in the morning in his parents' stone house, in the first-floor bedroom. It was the custom of the inhabitants of Sotto il Monte to name their houses, and the Roncallis called theirs the *Palazzo,* or palace, but there was nothing quite so grand about the place, especially since six cows shared the house with an extended family of Angelo's aunts, uncles, grandparents, and cousins. Angelo's birth brought the number of inhabitants in the Palazzo up to thirty-two.

Angelo was Giovanni and Marianna's fourth child. The fact that here was a boy, after three girls, caused great rejoicing for Giovanni, a tenant farmer who would now have someone to help him till his five hectares of land. But it was no less a cause of joy for Uncle Zaverio Roncalli, Angelo's great-uncle, the patriarch of the family, known to one and all as "Barba."

Shortly after Angelo was born, Zaverio and Giovanni and Marianna, who had roused herself from her bed and wrapped up the infant in shawls, went in search of Father Francesco Rebuzzini in order to have Angelo baptized. When they were told the parish priest was out, they simply sat down and waited on the cold benches of the church vestibule. There was no question of returning later—it was the custom of the family to baptize infants immediately, not only because of the high rate of infant mortality at the time, but because of the Roncallis' profound faith in their God and their Church.

Rebuzzini did not return until late that evening and stepped into the dark church to find Zaverio, Giovanni, and Marianna waiting for him. Although he was no doubt tired—tired enough to make a mistake about the infant's name in the church register—he knew that these devout people could not be put off, and he baptized Angelo in a simple ceremony, with Uncle Zaverio as the child's godfather.

And then the four returned home and the cycle of their lives began again—the cycle of faith, family, and farm. This cycle would give shape to the life of one of history's greatest popes.

After his election as supreme pontiff, Angelo Roncalli wrote, with typical humor, "There are three ways of ruining oneself—women, gambling and farming. My father chose the most boring." And the most arduous. Giovanni Roncalli was a sharecropper, tilling land that belonged to several wealthy landlords. He gave half his crop—wine, kale, milk and veal from the cows, silk from the silkworms on his cultivated mulberry bushes—to his landlords and kept the rest for himself. In good years, the Roncallis barely got by, but in bad ones they sometimes did not have enough to eat, given the mouths they had to feed: Giovanni and Marianna would go on to have ten more children after Angelo.

The Roncallis seldom even had bread, but made do with polenta (a dish made of corn flour). Despite this, they always had room at the table for one more. As Roncalli was later to write,

"When a beggar appeared at the door of our kitchen, when the children—20 of them—were waiting impatiently for their *minestrata*—there was always room for him and my mother would hasten to seat the stranger alongside us."

Roncalli became pope as the world sped into the 1960s, but he grew up in a place where time stood still. Roncallis had lived in Sotto il Monte since the early fifteenth century and were as much a part of the countryside as the stunted trees and the hills. (Their very name comes from the Italian word *ronchi,* an earthen terrace for planting grapes on hillsides.) They divided their days with church bells: The Angelus rang three times, awakening them at five in the morning; reminding Marianna to make lunch at noon; and calling the men in from the fields at six. The year was marked by Catholic feast days.

Angelo Roncalli and his brothers and sisters were surrounded by God and Church. When his Uncle Zaverio woke him up in the morning, the older man would say, "Time to get up, Angelo," and recite from the Angelus: " 'The Angel of the Lord declared unto Mary.' "

And Roncalli would respond, " 'And she conceived by the Holy Spirit. Hail Mary, full of grace. . . .' "

A short, stocky-legged, and powerfully built boy, Angelo worked in the fields alongside his father and the other Roncalli men, but he was especially close to Marianna, who provided him with his earliest memory: a journey to a shrine to the Virgin Mary, about a mile from the Roncalli home. On November 21, 1885, the feast of Mary's Presentation at the Temple, a great crowd gathered at the doors of the shrine. Unable to get her son inside, Marianna held him high to look through the window and said, "Look, Angelino, look how beautiful the Madonna is. I have consecrated you wholly to her."

With this as what he called "the first clear memory that I have of my childhood," it's no wonder that Roncalli would later write, "I can't remember a time when I did not want to serve God as a priest."

In October 1887, Angelo started at the village's only school, a simple one-room building with three benches, one for each grade. His younger brother Zaverio marveled that Angelo actually "wanted to go to school," and it showed. Angelo was quick and clever, sometimes too much so for his classmates. One day a visiting inspector of schools posed a trick question: "Which is heavier, a quintal of iron or a quintal of straw?" and all the children, quite sure of themselves, shouted out, "Iron!" There followed a silence, and then Angelo gave the correct answer: "A quintal is a quintal. They weigh the same."

Not taking too kindly to this, Angelo's fellow students beat him up. As one of them later remembered gleefully, "We waited for him on the street and we beat him up—just a little."

This didn't stop Angelo, who quickly became the best student in the village as he made his way through his primary school's three grades, taught in part by Father Rebuzzini. Angelo was a voracious reader with a near-photographic memory. When not in school, he either studied independently or under the religious tutelage of Uncle Zaverio. Barba, in his sixties and unmarried, was a profoundly faithful, well-read Christian who traveled to Rome in 1888 to attend the fiftieth anniversary of Pope Leo XIII's ordination as a priest.

On February 13 the following year, when he was eight, Angelo received the sacrament of confirmation at Carvico, the neighboring parish, by Bishop Gaetano Camilo Guindani. And two weeks later, on March 3, he made his first Holy Communion, a rare privilege for a boy of that age. (In the nineteenth century, children were generally not allowed the sacrament of the Eucharist until they were between ten and fourteen years old. It wasn't until the reign of Pope Pius X, twenty years later, that the age was changed to seven or eight.)

Later, as a young adult, Roncalli noted the occasion in his journal: "I was allowed to make my first communion on a cold Lenten morning, in the Church of Santa Maria di Brusicco. Only the children, the parish priest, Rebuzzini, and his curate, Don Bortolo Locatelli, were present." Roncalli remembered that Father Rebuzzini had asked the future pope to inscribe the names of the new communicants in the Apostleship of Prayer, which he said was "the first writing exercise I can remember doing, the first page of so many that would proliferate in half a century of living pen in hand."

The Apostleship of Prayer was a group of devout Catholics who daily offered their work and prayers to some good cause or intention approved by Pope Leo XIII. It was just another way for Roncalli and those who were as pious as he to devote every minute of their waking lives to Christ. The extraordinary thing about Roncalli, however, was that despite his evident religious zeal and sanctity, even as a very young boy, he was human—full of life and mischief.

While he respected and loved elders like Barba and others in the Roncalli family, he also saw through them. Half a century after he left his little village, he wrote to his brother Giovanni, "Fortunately, you brothers do not imitate our old people . . . who hardly ever spoke to each other except to grumble. . . . I remember that when I was a child I used to implore the Lord most fervently to make the old Roncallis talk to each other a little. And I used to wonder: how will they ever get to Heaven if the Lord says we must all love each other."

Despite their flaws, Roncalli loved his family deeply. He was not as close to his father, Giovanni, as he was to his mother, Marianna, but he carried with him deep in his memory a day in the summer after his first communion when his father held him on his shoulders to watch a religious parade in a nearby village. Years later, in 1958, on first being carried in the *sede gestatoria,* the portable papal chair, he recalled, "Once again I am being carried aloft by my sons. More than 70 years ago I was carried on the shoulders of my father. . . . The secret of everything is to let oneself be carried by God, and so to carry Him [to others]."

Roncalli began carrying God to others when he entered the Bergamo junior seminary in 1892, when he was nearly eleven years old. Entering the seminary at this time did not mean he was certain to become a priest—it was a kind of high school, the only place where gifted male children in the district could receive a higher education. Angelo knew in his heart that he had a vocation, but his father was far from convinced: "He is a poor farmer's son," Roncalli the elder said. "He'll make a poor priest."

Before Angelo left for school, his mother had collected all the money she could from the cash-strapped Roncallis, who lived mainly by barter, and presented it to her son. It was only two lire—a few cents—and she wept at how paltry it seemed, the only time the son could remember her crying.

Bergamo was only about 8 miles from Sotto il Monte, but it was in many ways situated in another universe. It was the biggest town in the area, had a shopping district and fine cafés, and was famous for its *commedia dell'arte* productions. The town's inner city, where the seminary was located, dated back to the time of the Etruscans.

Here Roncalli was exposed to a level of urbanity and sophistication he had never before experienced, but he also became aware of some of the harsh realities of the world beyond his home village.

Twenty-two years earlier, in 1870, Pius IX famously declared himself a "prisoner of the Vatican" after King Victor Emmanuel annexed Rome and the papal states. Pius IX, who ruled as pontiff from 1846 to 1878, forbade Catholics to hold national office or even to vote in the face of a resurgent Italian nationalism he deemed inimical to the Church.

This isolationism, however, could not survive the industrial revolution sweeping the world and Italy. Lured by factory jobs, workers left the farms and found their way to urban centers, where they worked grinding hours for pennies and lived in desolate conditions in slums—many such workers' ghettos were located right in the heart of the shining town of Bergamo. With both unemployment and taxes high, young Italians were leaving the country for France, Latin America, and the United States in record numbers—196,000 in 1888 alone. Both lay Catholics and clergy knew that the Church would now have to minister to a changing world.

In 1891, Leo XIII promulgated his famous encyclical *Rerum novarum* (The Condition of Labor), which began, "Some remedy must be found, and quickly found, for the misery and wretchedness which press so heavily at this moment on the huge majority of the very poor."

As the Church and the world began to change, Roncalli changed with it. In this regard, Bergamo was the place to be. Taking *Rerum novarum* to heart, Catholics, both clergy and laymen, set up soup kitchens, fought for the rights of the working class, and organized unions—by 1895, there would be nearly 100 unions or cooperatives representing 42,000 Catholic laborers and peasants, the very heart and soul of the Catholic Action movement in Italy.

For the time being, this fervor was out of reach for young Roncalli, who found himself struggling with subjects—especially science and mathematics—and with his good-natured inability to stop joking around in class and start paying attention. However, in his first few years in the junior seminary, he excelled at the things that really interested him—theology and history—and his grades improved.

When he entered the senior seminary in 1895, the fourteen-year-old Roncalli began to keep a journal, under the advice of the school's spiritual director, Canon Luigi Isacchi. The journal, kept in a series of black academic notebooks with stiff covers, was a way to stay in touch with his spiritual goals and to castigate himself when he went off course, as he did—by his own reckoning—with increasing frequency.

He would keep the journal for the rest of his life, ending up with thirty-eight notebooks and folders, which were published after his death in the volume titled *Journal of a Soul.*

"It seems quite impossible," he wrote in a typical entry. "The more I make resolves, the less I keep them. This is all I am good for: gossiping away, promising the earth and then? Nothing! If only I knew how to be humble.

"Sometimes I spend far too much time talking with the curate and it might be said of me, 'When words abound, sin is found.' Then there is another thing—I am very greedy about fruit. I must beware, I must watch myself."

Despite its trivial concern with gossip and diets, young Roncalli's journal carefully records the journey of a man on his way to becoming a priest. The first page of the first journal begins with a quotation from Ecclesiastics 3.27—"It is good for a man to bear the yoke in his youth"—and goes on to list the rules by which he hoped to live, including the time that should be spent each day in praying, attending Mass, meditating, reciting the Rosary, and, especially, in self-examination.

"Make a habit of frequently raising your mind to God," the young Roncalli writes at the same time as he admonishes himself to never "play or jest" with women, although he confessed that he had "two eyes in my head which want to look at more than they should."

Roncalli's early journals are aspirational—they record what he wished to be, and only sometimes was. Much later, reading over his notebooks as pope, he recalled, "I was a good boy, innocent, somewhat timid. . . . I imposed severe sacrifices on myself. I took everything very seriously."

But he loved Bergamo, and he would look back with great nostalgia and fondness on his eight years in junior and senior seminary there. He took long walks throughout the town, wearing the cassock and round hat of the seminarian. Although, as a pious student, he was not supposed to notice *too* much (he needed to keep "custody of his eyes," as the saying went), he saw both the ancient beauty of the town as well as the brawling taverns, the gambling houses, and the terrible poverty.

He did not yet know it, but he was receiving his education as a humanist—or perhaps, what he saw brought his naturally humanistic tendencies to the fore. Roncalli would not be a priest—or later, a pope—isolated from the modern world.

Perhaps not surprisingly—but quite painfully for Roncalli—he began to have problems when he returned home to Sotto il Monte. To his surprise, the elders of the village began to defer to him in embarrassing ways, as if he were already a priest. His cousins addressed him formally (*voi*) rather than intimately (*tu*), yet at the same time, it seemed to him, they considered him arrogant.

Perhaps, from their point of view, he was. In 1893, the family moved to a somewhat larger house, but the overcrowding and the continual squabbling there grated on Roncalli. "These cursed holidays," he wrote in his journal. "I have had three days of my vacation, and already I am tired of it. At the sight of so much poverty, in the midst of such suspicions, oppressed by so many anxieties, I often sigh and am sometimes driven to tears."

In June 1898, Roncalli returned home for a visit but grew frustrated by the continual family arguments. He was also upset because he was unable to find ink with which to write. A visiting Franciscan priest stopped by the house and heard tales— from Roncalli cousins, or perhaps from a disgruntled brother or sister—that Roncalli had demanded and received preferential treatment from his mother, getting more food than the rest of the family. The priest promptly took these rumors back to Roncalli's superiors at the seminary, who sternly rebuked him, forcing him, as he wrote in his journal, "to humble myself against my will."

Knowing that someone in his family had slandered him did not exactly endear the rest of the Roncallis to Angelo. In September of that year, he returned home again. On Saturday evening, September 24, he paid a visit to Father Rebuzzini, the parish priest who had baptized and, later, taught him. The two exchanged happy greetings, and the next morning, Roncalli saw him at Mass. But later that morning, Father Rebuzzini collapsed and died as he prepared for the liturgy. As Roncalli records in his journal:

> I did not weep, but inside me I turned to stone. To see him there on the ground, in that state, with his mouth open and red with blood, with his eyes closed, I thought he looked

to me—oh, I shall always remember that sight—he looked to me like a statue of the dead Jesus, taken down from the cross. And he spoke to me no more, looked at me no more.

Roncalli grieved deeply, writing in his journal, "I am left an orphan to my immense loss. . . . If my father has gone, Jesus is still here and opens his arms to me, inviting me to go to him for consolation." He was sixteen years old, and already he was proclaiming that his true family was the Church.

Taking with him Father Rebuzzini's copy of *Imitation of Christ* by Thomas à Kempis—Roncalli would keep it all his life—he returned to Bergamo. There was, of course, no longer any doubt that Angelo would become a priest. He had been given his first tonsure in 1895 and had just received the minor orders of lector and porter on July 3, 1898. A year later, on June 25, 1899, he received the minor holy orders of acolyte and exorcist. Even his studies had improved, and he was made prefect of his dormitory.

A year after Father Rebuzzini's death, while visiting a parish near Bergamo, Roncalli made the acquaintance of Monsignor Giacomo Maria Radini-Tedeschi, a canon of Saint Peter's Basilica in Rome. Radini-Tedeschi was a man of intelligence, power, and sophistication. An intimate of the elderly Pope Leo, he was spoken about as a future papal secretary of state, a liberal whose *Opera dei Congressi* was the oversight organization for Catholic social action groups that had sprung up across the country. Radini-Tedeschi told the young seminarian that if he ever came to Rome, he could study in his action group, Our Lady's Club.

Going to the great city of Rome seemed an unlikely occurrence to Roncalli, but then he had a stroke of extremely good fortune. Back in Bergamo, he was invited to sit for an exam for entry into the famed Pontifical Roman Seminary, the Apollinare, on a scholarship from the Flaminio Cerasola Foundation. Roncalli passed the exam and, along with two other Bergamo seminarians, took the overnight train to Rome, arriving at 6:30 A.M. on January 4, 1901.

The Apollinare turned out to be housed in a dark and ancient building located in a maze of twisting streets. Roncalli had a tiny room to himself. Although the mattress was hard, he had his own radiator, as well as electricity and running water, an almost luxurious existence for a young man from the country. "I could never have imagined," he wrote to his mother, "I would be so fortunate."

He set out to explore the Eternal City, the birthplace of Western civilization and ancient seat of Saint Peter the Apostle. He had arrived during a Holy Year, when pilgrims had made their way from across the globe to pay homage to ninety-year-old Leo XIII. Visiting the Propaganda Fide, the missionary college, Angelo was impressed as "forty clerics recited their own compositions in 40 different languages. . . . Some were white, some yellow, some red, and some had hands and faces as black as coal."

Roncalli painted a colorful picture of his teachers at the college, as well. There was Don Francesco Pitocchi, a priest with a rare disease that prevented him from lifting his head from his chest— yet, as Roncalli wrote, "He read our eyes; he read our hearts," and he acted as a profound spiritual adviser to the young seminarians. Roncalli's professor of Catholic Church history, Monsignor Umberto Benigni, published a Church newspaper and would often whisper to the seminarians summaries of the day's news, since they were forbidden to read newspapers themselves. Benigni, unfortunately, would later become an archconservative with anti-Semitic views, but at the time, Roncalli considered him a powerful teacher of history, a subject that had always fascinated the young seminarian—and would continue to occupy his thoughts in future years.

Another character was the bursar, Don Ignazio Garroni, from whom the penniless Roncalli was forced to borrow small sums of money. Garroni often strode through the dining hall during meals, urging the seminarians to "Eat less, eat less!" There was even a priest who claimed to be able to levitate.

Roncalli learned something from all his teachers, including Father Eugenio Pacelli, the tall, ascetic young lecturer in canon law who would later become Pope Pius XII. In June 1901, Roncalli won honors in theology and a prize for a paper he wrote in Hebrew. He had become so fluent in Latin that he was able to delve into medieval manuscripts at will. Life in Rome agreed with young Roncalli—but then his world came crashing down. The Italian government, still anticlerical, drafted him and many of his classmates into the army, refusing to make an exception for seminarians.

On November 30, 1901, he reported to Bergamo as a private in the 73rd Infantry Regiment, the Lombardy Brigade, for twelve months of service. Thus began what he later called his "year of Babylonian captivity."

Having his studies interrupted was part of what made the army so painful for Roncalli; the other part, as a fairly tender twenty-one-year-old who had been sheltered in seminaries since the age of twelve, was seeing how vulgar and sexually active his fellow soldiers were. "The army," he wrote in his journal, "is a running fountain of pollution."

But, in fact, Roncalli didn't do badly for himself, even in this culture. As a mountain boy used to walking, long marches held no fear for him, and he excelled, somewhat unexpectedly, on the rifle range. Before his year was up, he was promoted to corporal, and then sergeant, and returned to the Apollinare with real-world experience that would stand him in good stead when World War I broke out.

Pope Leo XIII died on July 20, 1903. He had reigned for twenty-five years, the second-longest-serving pontiff in history (until John Paul II a century later) and the only pope Roncalli had ever known. Leo had been a formidable force for progressive change within the Church. The favorite to replace him was Cardinal Mariano Rampolla del Tindaro, forty-nine, the Vatican secretary of state who would likely carry on Leo's liberal policies. In fact Rampolla led in the early voting at the conclave, but when the white smoke poured out of the chimney above the Sistine Chapel, the new pope was Cardinal Giuseppe Melchior Sarto, the patriarch of Venice. Why the sudden reversal?

Emperor Franz Joseph of Austria-Hungary vetoed Rampolla's candidacy in the last exercise of the *jus exclusivae,* or right of exclusion, traditionally claimed by Spain, France, and Austria since the sixteenth century, during which the secular rulers could eliminate papal candidates they found unpalatable.

Sarto, the newfound darling of the large bloc of conservatives, took the name Pius X. His election would have a profound influence on Roncalli's future. (And Roncalli would be the second of three patriarchs of Venice elevated to the papacy in the twentieth century.)

Roncalli was ordained a deacon on December 18, 1903. He graduated from the Pontifical Roman Seminary eight months later, in July 1904. On the morning of August 10, 1904, he was ordained a priest by Bishop Giuseppe Caeppetelli, the titular patriarch of Constantinople, in the church of Santa Maria in Rome's Piazza del Popolo. His parents and his Uncle Zaverio were unable to be there to see him reach this treasured goal—they couldn't afford the price of a train ride—but he wrote them immediately, filled with gratitude, and thanked them.

The next morning, Father Roncalli celebrated his first Mass in the crypt of Saint Peter's (where he would be buried some sixty years later, before his body was moved "upstairs") and then was brought to an audience with the new pope by Father Domenico Spolverini, the vice rector of the Apollinare. Father Spolverini told Pius, "Your Holiness, here is a young priest of Bergamo, who has just celebrated his first Mass." After congratulating him, Pius asked the youthful Roncalli—still three months shy of his twenty-third birthday, the canonical age for ordination—when he would be back home.

"For the Feast of the Assumption [August 15]," Roncalli replied.

"Ah, what a feast that will be, up there in your little hamlet," Pius X replied. "And how those fine Bergamesque bells will peal out on that day!"

And so they would.

CHAPTER THREE

Early Priesthood and Rome (1904–15)

O n the 15th [of August], the feast of the Assumption," the newly ordained Father Roncalli wrote in his journal, "I was at Sotto il Monte. I count that day among the happiest of my life, for me, for my relations and benefactors, for everyone."

The day was as joyous as Pius X had predicted. Roncalli celebrated Mass in the same church he had been baptized in twenty-three years earlier. He preached a sermon on the Assumption of the Virgin Mary. Many of those who had known him since his birth began to weep, at which point Roncalli interjected, "Dear brothers, my dear real brothers, seeing you cry this way unsettles me, although I know they are tears of joy."

But Roncalli did not stay around Sotto Il Monte for too long. Two weeks later, he left for Roccantica, a town about 50 miles north of Rome, where he spent time with recently ordained priests.

In November, he was back at the Apollinare studying canon law and working as a prefect for the incoming freshmen seminarians. He was busy and happy and idealistic, writing in his journal "that in all things there must be humility, great spiritual fervor, mildness and courtesy towards everyone." But his young priesthood was not without a few glitches. His spiritual adviser, Father Francesco Pitocchi, convinced him to give a talk on the Feast of the Immaculate Conception to the Children of Mary, a group of girls and young women who met in a chapel near Roccantica. Not used to the company of such privileged young women, he wrote out a flowery speech and then grew flustered as he made it:

> My talk was a disaster. I mixed up quotations from the Old
> and the New Testaments. I confused Saint Alphonsus with
> Saint Bernard. I mistook writings of the Fathers for writings
> of the prophets. A fiasco. I was so ashamed that afterwards
> I fell into the arms of Father Francesco and confessed my
> mortification.

His family, in particular his mother, was another and more serious problem. She wrote to him complaining that he wasn't sending her any money, that he had stopped caring for her, that he had sent a photograph of himself to the local priest rather than to her. He replied as patiently as he could. He had little money, he told her. As to the picture, he answered, "It is true that I have not yet sent a portrait to my family whereas I have sent it to the parish priest, but this is first of all because you already had it and he hadn't—and then I didn't want to distribute too many of my portraits, not being a pope or cardinal or bishop, but a mere humble priest."

But Roncalli soon faced bigger problems. The rise of Pius X to the papacy had led the Church away from the more liberal policies of Leo XIII and his advisers, including Monsignor Giacomo Maria Radini-Tedeschi, a canon of Saint Peter's Basilica, whom Roncalli met briefly before earning his scholarship to Rome. In July 1904, to the disappointment of many within the Catholic Church, Pope Pius dissolved the *Opera dei Congressi.* As chaplain of the organization, which promoted social justice, Radini-Tedeschi called Pius's actions "the hardest moment of my life."

In October, Bishop Gaetano Camillo Guindani of Bergamo died, and Pius appointed Radini-Tedeschi to the bishopric—a great honor, but also a way of moving him out of the centers of power in Rome. The bishop-elect needed a secretary and apparently tried out Roncalli and his young Bergamo friend and fellow priest Guglielmo Carozzi for a week each before settling on Roncalli. On January 19, 1905, as Pius ordained Radini-Tedeschi bishop, Roncalli held the Book of Gospels over his shoulders, signifying the burdens his mentor would carry as the new bishop of Bergamo.

Then Bishop Radini-Tedeschi departed for Bergamo, taking his new secretary with him. It was the beginning of an extraordinary relationship between Roncalli and the man he called "my spiritual father" and "the Pole Star of my priesthood."

Giacomo Maria Radini-Tedeschi was born in 1857 in Piacenza, a farming community about 40 miles south of Bergamo, but his early life was far different from Roncalli's. His family was an ancient and noble one; he was actually a count, but he surrendered his title when he became a priest.

A firm believer in social justice and Catholicism as a spiritual force that could change the lives of the poor and less fortunate,

Radini-Tedeschi rose quickly through Leo XIII's Church. His new assignment seemingly put an end to his ascendancy, but he accepted his post with a humility that did not go unobserved by his new secretary.

He and Roncalli arrived in Bergamo on April 19, 1905. The pair must have made quite a picture. Radini-Tedeschi—"so tall and noble in person and manner," as Angelo was later to write— was as thin and ascetic-looking as Roncalli was short and squat. If one were casting a count and a peasant in a play, they would be ideally suited for their roles. The effect was heightened by the fact that the bishop's carriage fit only one seat, forcing Roncalli to run behind it. "I willingly puffed along," he later wrote, "but the poor bishop never enjoyed the ride, for he kept worrying about me and looking back to see that I hadn't collapsed." The experience led Roncalli, as pope, to add a seat in his official car for his secretary.

Radini-Tedeschi was a brilliant, nervous, autocratic multi-tasker who paid attention to the slightest detail and could occasionally come across as a martinet. Roncalli was a very different personality, but the two men, bonded by their enormous energy, complemented each other. They had barely gotten settled into the episcopal palace in Bergamo when Radini-Tedeschi headed off for the tomb of Saint Charles Borromeo in Milan, and then to the shrine at Lourdes. Roncalli accompanied him. The men would make numerous such pilgrimages together, trips that were invaluable in broadening Roncalli's outlook on the Church in other parts of the world.

After their initial trip, Radini-Tedeschi and Roncalli—soon to be known as "the Bishop's shadow"—set about transforming Bergamo. In 1905, there were 430,000 Catholics in the diocese, with 350 parishes, 512 churches, and 2,000 priests, all of which fell under the bishop's responsibility. And the place, it seemed, was falling apart. The episcopal palace was, according to Roncalli, "an ugly, inconvenient and insanitary building," so Radini-Tedeschi ordered a new one built and moved into it in 1906. The seminary was without running water or electricity, and Radini-Tedeschi modernized it. The seventeenth-century cathedral was extensively renovated, and new schools and churches were built throughout the diocese. Radini-Tedeschi began a series of pastoral visits to every parish; Roncalli estimated that the bishop had personally given communion to at least a third of all of Bergamo's Catholics.

Radini-Tedeschi's energy and attention to detail won over the clerics and laypeople in Bergamo who had assumed Radini-Tedeschi would arrive from Rome with an attitude that their northern diocese was beneath him. For his part, it became clear that, over and above his construction projects, Radini-Tedeschi intended to continue to carry on the principles of Catholic Action that he had so treasured. As Roncalli would later write in his biography of Radini-Tedeschi, "He grieved for the disappearance of [the *Opera dei Congressi*], but always remained faithful to its ideals . . . [and wanted] to direct its rejuvenated energies into new social organizations required by the new conditions of the times."

For his part, Roncalli's education with Radini-Tedeschi taught him that there was a way to apply changes within the Church while preserving older traditions, an important lesson he would apply later during the Second Vatican Council.

Through Radini-Tedeschi, Roncalli met a fascinating group of liberal clergy who sought to fulfill their traditional roles of administering to the faithful while pushing forward the idea of Christ as an instrument of social change. "Prudence," Radini-Tedeschi told Roncalli, "does not consist of doing nothing. It means to act, and act well." Radini-Tedeschi organized a travel aid service to help the workers who were forced by the Italian government to immigrate to other countries. He established three organizations to help women: the League of Women Workers, which protected women in the workplace at a time when they had no rights whatsoever; the Association for the Protection of Young Women; and the *Casa di Maternita,* which aided expectant mothers. (Roncalli served as adviser to these three groups.)

In 1909, when iron workers went on strike in Ranica, a small town outside Bergamo, Radini-Tedeschi made a personal contribution to the workers' fund, and he and Roncalli visited their homes with food and clothing. As Roncalli later wrote, "The question was not a simple one of wages and personalities, but one of principle: the fundamental right of Christian labor to organize against the powerful organization of capital."

Not everyone accepted this, naturally—"Less than benevolent reports were dispatched to our superiors in Rome," Roncalli wrote dryly—but Radini-Tedeschi was able to convince Pope Pius that his work with the strikers was justified. Radini-Tedeschi wrote directly to Pius, who responded, "We cannot disapprove of what you have thought prudent to do, because you are fully acquainted with the place, the persons involved and the circumstances."

In September 1905, through Radini-Tedeschi, Roncalli met Cardinal Andrea Carlo Ferrari, the archbishop of Milan. Despite their twenty-year difference in age, the two men became fast friends. Ferrari also served as Roncalli's spiritual adviser. Ferrari was out of favor with Pius X for his liberal views—in fact, one historian has suggested that Pius orchestrated a slanderous campaign against Ferrari, whom Roncalli, as pope, would beatify in 1963. It was typical of both Roncalli and Radini-Tedeschi that, while it was not politically expedient to be a friend of Ferrari, they continued to openly visit and consult him in Milan.

One such visit, in February 1906, bore important fruit. While Radini-Tedeschi and Ferrari discussed a forthcoming clerical meeting, Roncalli browsed idly in the cardinal's library, which included ancient tomes and dusty parchments. There he found something extraordinary, as he later wrote: "Suddenly I was struck by 39 bound parchment volumes which bore the title: *Archivio Spirituale—Bergamo*." What Roncalli had discovered were documents that described the pastoral visits of Saint Charles Borromeo to the Bergamo diocese in 1575, a detailed account of the Church life in the Renaissance.

Determined to edit and publish these documents, he returned to Milan again and again over the course of 1906 to work. In order to aid his secretary, Radini-Tedeschi introduced Roncalli to the prefect of Milan's Ambrosian Library, Monsignor Achille Ratti, who later became Pope Pius XI. Ratti agreed to help Roncalli prepare the documents for publication. The first volume of *Record of the Apostolic Visit of Saint Charles Borromeo to Bergamo, 1575,* came out in 1936. The last appeared in 1958. As Roncalli wrote wryly, "As so often happens, so it happened in this case too: A project begins with the naming of committees, but the work has to be done by a single person."

Roncalli was now gaining a reputation. More and more, he was viewed as an indefatigable activist: "Our fine Don Roncalli has tried to organize even the telephone operators," a Bergamo businessman noted ruefully. "Would that he were satisfied just to organize the sacristans."

But he was not. His working education with Radini-Tedeschi was indeed teaching him the importance of prudence, in the true sense of the word: "to act, and to act well." Radini-Tedeschi encouraged Roncalli to spread his wings. He knew his young charge was no mere secretary, but a man in possession of many talents.

Roncalli spent a great deal of time filling in for priests who were absent in their parishes—saying Mass, visiting the sick, hearing confessions, teaching children catechism—so that he could understand the very beating heart of pastoral work.

In 1906, he began teaching church history and apologetics, the defense of the Catholic faith, at the seminary in Bergamo. His students later remembered him hurrying into the classroom, late and out of breath, entertaining them with an anecdote or two, and then displaying an unexpected breadth of knowledge for someone so young. "Always be prepared to answer someone who demands a reason for your faith," he told his students.

Throughout the ten years he served as Radini-Tedeschi's secretary, Roncalli continued to travel. Between 1905 and 1913, he and Radini-Tedeschi made five pilgrimages to Lourdes, attended numerous conferences all over Europe, and also made a pilgrimage to the Holy Land in 1906. Roncalli wrote about the latter trip, sending dispatches back to *L'Eco di Bergamo* that showed a good deal of writing talent. Here he describes an early morning trip across the Lake of Tiberias:

> I shall never forget the enchantment, the heart's ease, the
> spiritual relish I discovered this morning floating upon
> these waters. Little by little, as our small boat stood out into
> the lake, the first light of dawn lent color to the water, the
> houses and then the surrounding hills. We did not speak,
> but our hearts were stirred. It was as though we could see
> Jesus crossing this lake in Peter's boat. Jesus was before us
> and we could see him; unworthy though we were, we sailed
> towards him and our prayer, silent though it was, was eloquent and spontaneous.

Seeing the Holy Land, where Christ lived and walked, was inspiring for Roncalli. It gave him a feeling of the purity and essence of Christianity—a feeling the previous passage distills movingly and poetically: "We sailed towards him." They approached the living Christ both literally and figuratively.

There was nothing quite so pure about the wars of faith going on in the Church, however. Pope Pius's 1907 encyclical *Pascendi Dominci gregus* (Feeding the Lord's Flock) was a thundering attack on "modernism," which was then vaguely defined as a rational approach to the Bible, the separation of the Christ of history and the Christ of faith, and the belief that dogma was mutable and could, therefore, change over time.

Pascendi heralded what amounted to a crackdown on modernist or humanist belief within the clergy. Radini-Tedeschi, of course, was a prime proponent of humanism. Councils of vigilance were set up within dioceses to watch for any deviation from Church doctrine. Umberto Benigni, Roncalli's old professor, was encouraged by the Vatican to set up a secret network of spies and informers. In late 1907, the Vatican let it be known that anyone who disagreed with *Pascendi* would be excommunicated, something that made Roncalli's teaching of Church history problematic, given his sympathies.

Between 1907 and Pius X's death in 1914 there ensued what one bishop privately called a "white terror" in which "the Church [desired] not only to tell us what must be believed, but *how we should think.*" Roncalli tread cautiously in his lectures, allowing that humanistic thinkers had made certain "errors," while simultaneously referring to "narrow-minded and ancient formulations that had lost all meaning" that the Church was said—by *some,* Roncalli was careful to add—to be locked into.

In the meantime, his old world, the world of Sotto il Monte, was fast falling behind. In May 1912, he learned his beloved Uncle Zaverio was ill. Roncalli arrived back home the day before Zaverio died, at the age of eighty-eight. He wrote the memorial card, which seemed to yearn for a simpler time and place when basic virtues were rewarded: "He was the just man of Sacred Scripture. Simple, honest, God fearing, humble of birth, he had a lively and profound *sense of Christ*. . . . In a century full of agitation he never lost his youthful, fervent and loving devotion to the Sacred Heart of Jesus."

This period in Roncalli's life gradually drew to an end, with both painful and welcome consequences. The most painful was the psychic and physical distress of his greatest mentor, Radini-Tedeschi. By the summer of 1914, it had become increasingly evident that he was quite ill. Radini-Tedeschi lay confined to his bed in a villa in the mountains outside Bergamo, gradually losing weight and growing weaker and weaker, even though he was only fifty-seven years old. The bishop was further troubled by the fact that he thought that Pope Pius X had turned against him for good.

Despite his disagreement with the pontiff on these matters, Radini-Tedeschi was an orthodox Catholic cleric for whom it was painful to be at odds with the spiritual leader of the Universal Church. He felt that he had been slandered by the spies and informers who had infiltrated his diocese. A few years before, he had written Pius, telling him, poignantly:

I wish to be a pastor and a father, to try to win the people
with affection, with much affection, without giving way
to weakness. Perhaps all this, if judged by someone who is
not enlightened by the Holy Spirit, or who does not feel the

pains of being father to all, might appear to be remissiveness, or excessive goodness, or an inclination to see everything in a rosy color. Rosy views and remissiveness were perhaps those sins, Holy Father, which brought upon me the accusations of being intransigent.

It is not known whether Pope Pius ever responded to this, but Bishop Radini-Tedeschi grew more despondent as he grew more and more ill. His depression was not helped any by the assassination of Archduke Franz Ferdinand and his wife in Sarajevo that June, which moved Europe inexorably toward war. In August, Radini-Tedeschi was taken to Bergamo for an exploratory operation; it turned out he had inoperable cancer of the intestines, which had spread throughout his entire system.

Radini-Tedeschi was brought back to his mountain villa to die. The bishop, as Roncalli later described in his biography of him, grew despondent, weeping uncontrollably and bemoaning the fact that he had not helped enough people. "I am a bishop," he told his secretary. "I am a bishop with great power to do good, and I have not done enough. And now God is to judge me."

On August 20, 1914, Pope Pius died in Rome of a heart attack, in great measure brought on by his horror at the war that had already begun. Two days later, on the evening of August 22, Roncalli prayed next to an increasingly feeble Radini-Tedeschi. Thinking the bishop had fallen into unconsciousness, he stopped.

"Courage, my dear Don Angelo," Radini-Tedeschi told his secretary. "It goes well. Continue, for I understand every word you say."

Just before midnight, he died.

• • •

In the weeks following the death of his beloved mentor, Roncalli felt adrift in the world. He was thirty-three years old and had spent his entire young priesthood in the service of his bishop. The gathering war clouds had burst over Europe, unleashing a bloody rain.

Pope Pius X was dead, and the fifty-seven cardinals (of sixty-five eligible electors) formally met in conclave to elect his successor on August 31, with voting beginning the next day. On the third day, after eight ballots, the cardinals were deadlocked, with Giacomo della Chiesa, the archbishop of Bologna and former official of the Secretariat of State, leading but not yet able to achieve the two-thirds majority required for election. Della Chiesa, a close friend of Radini-Tedeschi and an admirer of Pope Leo X, was fifty-nine years old and newly elevated to the red hat. The "conservatives" were united behind Domenico Serafini, age sixty-four, assessor of the Holy Office (the powerful doctrinal congregation), who had, like the front-runner, been named a cardinal by Pius X in the consistory of May 25, 1914.

On the morning of September 3, 1914, the deadlock was broken, and after two more scrutinies, or votes, the Bolognan was elected and took the name Benedict XV. At least he was a man whose leanings were more sympathetic to Roncalli's own, something that young Roncalli found heartening.

Shortly after Radini-Tedeschi's funeral, Roncalli moved out of the Bergamo episcopal palace and took up residence in an apartment with his old seminary classmate Guglielmo Carozzi. He continued to teach, to work with the women's groups that Radini-Tedeschi had established and that Roncalli helped oversee, and to edit the parchment manuscripts of Charles Borromeo. It was a time of questioning for him. In late September, he went on a weeklong retreat with the Priests of the Sacred Heart and spent time reflecting on his life in his journals:

O God, your purposes are unfathomable! Immediately
after [the tenth anniversary of Roncalli's ordination, on
August 10] . . . you called my revered Bishop to share your
heavenly joy, and here I am in an entirely new situation. . . .

I will endeavor not to feel any anxiety about my future.
I was born poor and I must and will die poor, sure that at
the right time Divine Providence, as in the past, will provide
what is needed, sending me what I require and even more.

As Roncalli continued with his writing and teaching duties in
Bergamo, however, historic events were catching up to him. Although there was a strong sentiment within many Italians that
their country should stay out of the war, the Italian government
was offered secret inducements by the Allies if it would aban-
don its treaty partnership with Austria-Hungary and Germany
and enter the fighting. This it decided to do, with disastrous
consequences. On May 23, 1915, Italy declared war on Austria-
Hungary, with whom it shared a border.

Even before that, Roncalli and thousands of Italian men received call-up notices. On the evening of May 23, Angelo seized
by both fear and faith, wrote in his journal:

Tomorrow I leave to take up my military service in the Medical Corps. Where will they send me? To the front, perhaps?
Shall I ever return to Bergamo, or has the Lord decreed that
my last hour shall be on the battlefield? I know nothing; all
I want is the will of God in all things and at all times, and
to work for his glory in total self-sacrifice. In this way, and
in this way only, can I be true to my vocation and show in
my actions my real love for my country, and the souls of my
fellows.

CHAPTER FOUR

The Great War and After (1915–25)

Angelo Roncalli made an unusual soldier, to say the least. He was even stouter than the first time he had entered the army as a young seminarian in 1901. His countenance in photos taken at the time is one of serenity and benevolence, not martial determination. He grew and carefully groomed a moustache—"a weak moment on my part," as he later put it—but it didn't help. He was who he was, a man of God.

After reporting for duty in Milan, he was given his old rank of sergeant, assigned to the medical corps, and then sent back to Bergamo to work in the hospitals there. During the next three years, the Italian Army would attack the Austrians along a front in northern Italy about 200 miles north of Bergamo. By war's end, Italy had suffered 66,000 dead and 190,000 wounded. Bergamo was a major receiving center of these casualties, which placed it in the bloody center of the war.

Working feverishly in the hospital, Roncalli seldom slept more than five hours a night, working as both lowly medical orderly and priest. When there was nothing he could do physically for his patients, he spent his time ministering to them spiritually, as he was to write, "offering to the dying the last consolation of friendship and the reconciliation of final absolution."

As a priest, he was an unusual presence in his unit, especially for the anticlerical individuals among the officers and men. One lieutenant colonel dealt with him so rudely and sarcastically that Roncalli, afraid his men would begin to do the same, complained about the man. The officer returned with a sarcastic but strangely prescient apology: "I am only a poor lieutenant colonel," he said, "whereas you are only at the beginning and shall probably become a cardinal." At the very least, however, Roncalli was able to spend more time keeping track of his large family and offering his services to them in any way he could. Four of his brothers had been inducted into the army; a cousin had just been killed in combat. He was especially concerned about Zaverio and their youngest brother, Guiseppe.

After Zaverio passed his induction exam in September 1915, Roncalli wrote to him, "Keep up your spirits and your confidence in God; with God it is always good, whether at the front or not."

In March 1916, all the priests in the Italian Army were made chaplains and promoted to officer. Sergeant Roncalli was now Lieutenant Roncalli. He grew even busier. In addition to this work in the city's hospital—particularly its huge military reserve hospital, New Shelter—he was assigned to the civil defense department of Bergamo as a clerical adviser. He continued to teach at the seminary, wearing his clerical garb, although his sleeves and round hat carried two gold stripes to connote his military rank.

At the same time, he somehow found the energy to finish his biography of Radini-Tedeschi and even traveled to Rome to present a copy to Pope Benedict XV. In the introduction, he penned, "These pages were written while in Europe the war went on, the horrible war that caused so much bloodshed and tears. I have written these lines and worked on this book not in the sweet quietness of the life of studies but amidst the most varied occupations . . . first for several months as a simple soldier, then as a noncommissioned officer of the lowest rank, and finally more direct as a priest."

Roncalli's modest listing of the various roles he played during the war gives one some idea of his useful ability to adapt himself to new situations and to get along with the people he meets in them. The troops Roncalli was ministering to were, for the most part, simple peasants who knew little of the reasons for which they were fighting; they suffered with such great humility, Roncalli later wrote, "that I had to fall on my knees and cry like a child, alone in my room, unable to contain the emotion I felt."

While Roncalli's journal entries from wartime are mostly lost, one entry survives from March 1917. It captures the simple piety of the men he administered to, as well as his own:

How dear to me is Orazi Domenico as he struggles with a violent crisis of bronchial pneumonia in a room not far from mine. He's 19 and comes from Ascoli Pisceno. A humble peasant with a soul as limpid as an angel's. It shines out from his intelligent eyes and his good and ingenuous smile. This morning and evening, as I listened to him murmuring in my ear, I was deeply moved. [He said to me:] "For me, Father, to die now would be a blessing: I would willingly die because I feel that by the grace of God my soul still remains innocent. If I died when I were older, who knows how lumpish I'd become."

Despite Roncalli's prayers, Domenico died a month later. Roncalli mourned him greatly, not only for who the boy was but also for what his innocent piety represented. "The world needs such chosen, simple souls who are a fragrance of faith, of purity, of fresh and holy poetry," Roncalli reflected. "And we priests need them too to feel encouraged to virtue and zeal."

Roncalli's mourning for the shattered youth of Italy would increase in fall 1917, when the Austrians, bolstered by Russia's collapse on the eastern front and aided by seven crack German divisions, launched a massive offensive against the Italian Army at Caporetto. Within two weeks, they had taken 300,000 Italian soldiers prisoner and sent the rest reeling south, driving them to within 15 miles of Venice, a loss of 70 miles of territory. Forty-five thousand Italians were killed and wounded, many of them by poison gas, in a catastrophic defeat that left a million and a half civilians under Austrian rule. The Italians were able to hold on only with the aid of French and British reinforcements sent from France and Belgium.

The retreat was so headlong and ignominious that Italian commanders ordered those who fled decimated, in the old Roman tradition—that is, one out of every ten was shot as an example to the others. But it did not seem to stop the thousands of peasant Italian soldiers who were simply turning away from a war that was not of their own making and in some ways concerned them little.

Roncalli's life became an exhausting cycle of caring for the wounded who flooded Bergamo, overflowing the hospital facilities, lying on stretchers and crude pallets on church pews and in the hallways of public buildings. But he had a more personal reason for concern; he had sent a letter to his brother Guiseppe, only to have it returned "Addressee Missing in Action." Guiseppe, only twenty-three years old, had gone missing during the great battle. It wasn't until early 1918 that Roncalli discovered that he was safe, though wounded and ill, a prisoner of the Austrians.

By April 1918, the war had quieted on the Italian front, and Roncalli wrote, "I am almost without patients here." He began to turn his mind to other matters. Bishop Luigi Marella of Bergamo asked him to set up a student hostel, and he threw himself whole-heartedly in the project, a way to think hopefully ahead, to a time when the war would end. He found a building on the Via San Salvatore in the old city, refurbished it, and paid for its furnishings himself with a 2,000 lire loan (paying interest of 5 percent) and a gift from his father.

He invited his two unmarried sisters, Ancilla and Maria, who had previously kept house for him, to become the housemothers of the hostel.

The student hotel—*Casa dello Studente*—was not for seminarians, but for those lay students studying in Bergamo with nowhere to go. The thirty-seven-year-old Roncalli knew how to appeal to the young men. On the first floor he hung a full-length mirror over which he had lettered, in Latin, the ancient aphorism, "Know Thyself." When one unshorn student stared at it, wondering aloud what it meant, Father Roncalli walked up behind him and said, "You need a haircut."

The Casa dello Studente was a great success, but as the days progressed more sorrow came into Roncalli's life. In October, his sister Enrica died of cancer at the age of twenty-five. His brother Giuseppe Luigi was finally repatriated to Italy, although he remained seriously ill before eventually recovering. At the same time, a reorganized Italian government launched a new offensive against the Austrians to gain back the ground lost at Caporetto.

Finally, the war ended, on November 11, 1918. Roncalli could not have been happier. Officially released from the army in February 1919, he burned his uniform in farewell and celebration.

"During these last years," Roncalli wrote in his journal in April 1919,

> there have been days when I wondered what God would
> require of me after the war. Now there is no more cause
> for uncertainty, or for looking for something else: my main
> task is here, and here is my burden, the apostolate among
> students. . . . How true it is that when one entrusts oneself
> wholly to the Lord, one is provided with everything needful!

The period immediately following the war was a peaceful one for Roncalli, as well as for Italian society as a whole. As well as running the Casa dello Studente, he served as spiritual director of the Bergamo seminary. He also continued as spiritual adviser of his Catholic Action women's groups and organized the first postwar eucharistic congress, a dream of his mentor Radini-Tedeschi. It was held in Bergamo in 1920, and Roncalli spoke on "The Eucharist and Our Lady, Loves of the Christian." According to one of the attendees, Roncalli's speech was interrupted by applause numerous times and ended in a standing ovation.

Before that, in 1919, Roncalli visited Rome for a conference of the National Union of Catholic Women, hoping to stroll through some of his old Apollinare haunts, but things did not go well. His wallet and 800 lire were stolen at the train station in Milan, and he found the bustle of Rome tiresome. He did, however, get an audience with Benedict XV, although he had to spend all morning waiting for one ("This blessed audience," he later wrote in his journal, "is becoming a positive torture for me"). Little did Roncalli know this pope would soon change his life dramatically.

Although Benedict had been unsuccessful in a peace plan he had put forward in 1917 in an attempt to bring warring parties to the treaty table, he was extremely energetic in the years immediately following the war in reversing some of the isolationism the Church had felt during the papacy of Pius X. After Woodrow Wilson, whose Fourteen Point Plan echoed Pius's own peace plan, visited the Vatican, other countries sought diplomatic ties with the Holy See, and Pius encouraged Italian Catholics to return to politics within their own country.

Developing a Catholic political base was wise policy. Italian Socialists, inspired by their Russian counterparts, tangled with right-wing activists—soon to be known as Fascists, the precursor to Benito Mussolini's black shirts. Benedict was also concerned about the fate of the worldwide Church in developing countries, feeling the first stirrings of independence after years under the yoke of imperialism.

Benedict's November 30, 1919, encyclical *Maximum illud* (On the Propagation of the Catholic Faith Throughout the World) focused on the importance of missionary activity as the Church entered a new era. Since 1822, Catholic missions had been funded by an organization called *Propaganda Fide,* or the Society for the Propagation of the Faith. The society had started in France, but other, similar branches had grown up in numerous countries around the world, providing food, medicine, and catechisms to those the Church was attempting to convert and educate, particularly in Africa and South America.

These groups had done worthy work, but they were disorganized and too independent, and their approaches to missionary work were sometimes antiquated. Benedict XV wanted to put them under the central authority of the Church in Rome, bringing them together under the direction of a single Vatican department. He needed a man to do this for him—someone who could handle the delicate task of persuading wealthy clergy to share funds they had raised and laymen money they had earned for the common good of the missions.

Presented with a list of possible candidates in December 1920, Benedict apparently pointed to Roncalli's name and said, "That one, that one." Whether he remembered Roncalli because of his biography of Radini-Tedeschi or his work on the eucharistic congress, his desire was clear. Cardinal Willem van Rossum, the Dutch prefect of the Propaganda Fide—so powerful within the Church that he was known as the "Red Pope"—sent a letter to Bishop Marella of Bergamo requesting Roncalli for the job.

Marella passed the summons on to Roncalli, who found himself, as he would later write, "in perplexity and pain." Naturally, such a summons flattered the thirty-nine-year-old priest, but Roncalli didn't think he possessed the organizing ability, or the tact, for such a big job. Rather amusingly, he told Bishop Marella that he considered himself "someone who doesn't get much done; by nature lazy, I write very slowly and am easily distracted."

It did not work out as he anticipated. As a last resort, Roncalli wrote a letter to Cardinal Andrea Carlo Ferrari, the archbishop of Milan, who had continued on as his mentor after the death of Radini-Tedeschi. Ferrari, dying of throat cancer and barely able to speak, wrote him back on December 15, 1920: "You know how fond I am and, too, this is an obligation to Monsignor Radini. For this reason, here is my frank, unhesitant opinion. The will of God is perfectly clear. The Red Pope is the echo of the White Pope and the White Pope is the echo of God. Relinquish everything and go, and a great blessing will go with you."

And so he did. Arriving in Rome in January 1921, Roncalli was taken to a private audience with the pope, in the papal apartments on the fourth floor of the Apostolic Palace. Benedict told him that he was to be "God's traveler," heading for Propaganda Fide centers within Italy and in France, Belgium, and Germany, where his job would be to convince those who had long held power in the organization that they needed to let go of the reins of authority.

That spring, Benedict made Roncalli a monsignor, which made his family proud and gave him some cachet with the men he would be dealing with, but mainly he had to operate using his own instincts and the skills of persuasion he had learned firsthand in the company of prelates like Radini-Tedeschi and Cardinal Ferrari (Ferrari died on February 2, leaving Roncalli once again without a spiritual mentor).

Before his travels, he found a house in Rome where he could stay upon his return. His two sisters, Ancilla and Maria, who depended on him for their livelihood, agreed to look after the house in his absence. Concerned about his old seminary rector Monsignor Vincenzo Bugarini, who was ill and approaching seventy, Roncalli moved him into the house, as well.

Roncalli had little money—"Let us hope that Providence will help us make ends meet," he recorded—but, with a home for himself set up and those he cared about taken care of, he set off on his travels.

After visiting throughout Italy in the summer and fall, Roncalli headed to France, Belgium, Holland, and Germany, traveling from December 17 to January 8. On his trip, he learned exactly how each mission was funded and set up and how the

Vatican would eventually assume control. Shortly after Roncalli returned to Rome, the normally healthy Pope Benedict XV contracted pneumonia and died on January 22, 1922, at the age of sixty-seven.

There followed a historic conclave in which both progressive and conservative elements of the Church vied to have their candidates succeed Benedict.

The conclave began on Thursday, February 2. All but one of the fifty-three electors in Rome attended (he was ill with influenza), and two others were sick enough to miss the first two ballots, which revealed the division within the College of Cardinals. Each existing faction put forward candidates, promising another deadlocked process. By the time the cardinal electors were at full strength, with fifty-three electors in attendance, the blocs were clarified: twenty-eight liberals to twenty-five conservatives, with thirty-five votes required for election.

The conclave turned to Cardinal Achille Ratti, former librarian at the Ambrosiana in Milan and, later, at the Vatican, who had been made a cardinal by Benedict XV on June 13, 1921, and succeeded Ferrari as appointed archbishop of Milan. Ratti won forty-two votes on the fourteenth ballot on Monday, February 6. He took the name Pius XI.

Ratti was a brilliant, scholarly man, but, politically speaking, a bit of a cipher. Though widely assumed to be a moderate, no one knew for certain. This is one reason for his election. Regardless of his politics, however, he supported Roncalli's work for the missions.

When Roncalli and Cardinal van Rossum, the Red Pope, outlined the details of transferring authority of the society from the Propagation of Faith to the Vatican, the pope issued it under his own signature in March.

Despite this vote of confidence, Roncalli was a little leery of the politics of the position. In July of that year, he wrote a friend:

The new Holy Father is well. I saw him again in a long
audience a few days ago. He had the goodness to treat me
with the trust that befits an affectionate friend of Mgr.
Radini-Tedeschi and the prefect of the Ambrosian Library.
Yet I get between his feet as little as possible, and feel shivers
run down my spine every time I have to go through these
Vatican halls. Despite my constant and heart-felt attempt
to serve the new pope as best I can, I don't envy—indeed I
feel compassion—towards those who have to work in the
Vatican.

It was not distaste for the new pope that kept Roncalli away from the Vatican, but distaste for politics in general. But he continued to succeed. Between 1920 and 1922, he raised the society's collection from 400,000 lire to more than a million. Ever mindful of the power of the printed word, he started a magazine called *The Propagation of the Faith in the World* that told stories of the founding of the society and the labors of the faithful, both lay and clergy, throughout the world.

Yet he did all this with typical tact. He was always sure to praise the heads of local missionary societies. He also used his connections with Monsignor Bugarini, his housemate and the former seminary rector, to forge bonds with former Apollinare graduates throughout Italy. He had come to love his travels. Of one visit to Sicily, where he had never been, he wrote in glowing terms, "It is like a garden. . . . I can truly say it is a semi-paradise." And the food was so good it had given him "an extraordinary appetite."

As happened so many times in Roncalli's life, the occupation he thought he had established was replaced by another one. Italy of the early 1920s was a tumultuous place. Plagued by economic problems, high unemployment, and continual strikes, the country saw the violent and vocal Fascist groups, led by Mussolini, continue to grow. In 1922, Fascists attacked a Corpus Christi parade, shouting *Abasso il Papa* ("Down with the pope!"). Although they were still in the minority in Parliament, their violence threatened to bring down the government, and a weak King Victor Emmanuel III asked Mussolini—soon to be il Duce—to form a new government. In October 1922, Mussolini took power and began his decades-long dictatorship.

This was the beginning of the complicated and often unsavory relationship between the Italian Church and fascism. Though Mussolini was an atheist, he knew he needed the Church. "Since the Italian people is all but completely Catholic," he famously said, "and Catholicism is the ancient glory of Italy, the Italian nation can be nothing less than Catholic." In order to keep Catholics from rallying against him, he introduced religious education to public primary schools and even endorsed the return of crucifixes to public buildings, where they had not hung since the nineteenth century.

His gestures paid immediate dividends. In January 1923, in advance of the 1924 spring elections, Cardinal Pietro Gasparri, Vatican secretary of state, handed Mussolini a list of supporters of the Partito Popolare, the nascent political party that would eventually become the Christian Democrats, "unreliable and mistaken men" within the Vatican who could not be trusted.

Additionally, the meeting marked the beginning of the secret negotiations that would ultimately lead to the 1929 Treaty of the Lateran, in which the Holy Father renounced claims to the former Vatican states, and the Italian government recognized the inviolability and independence of what came to be known as Vatican City State, which still exists today as a tiny independent nation, with the pope as sovereign head of state.

If this level of collaboration on the part of the Vatican with Mussolini is shocking, it must be remembered that at this early stage Mussolini had numerous international supporters, among them Winston Churchill and Herbert Hoover. Still, in February, Roncalli sent his family a letter, telling them, "I recommend everyone not to get too excited about the elections. Vote when the time comes. Now it is better to let things be. Keep quiet and stay at home; think it out for yourselves." He, however, vowed that he would "remain faithful to the *Partito Popolare*." In another letter sent to them just before the elections, he spoke even more forcefully: "In my conscience as a priest and a Christian, I don't feel I can vote for the Fascists. . . . Of one thing I am certain: the salvation of Italy cannot come through Mussolini even though he may be a man of talent. His goals may perhaps be good and correct, but the means he takes to realize them are wicked."

Roncalli's comments about Mussolini's means were proven correct that election day, as his Fascists mobbed election booths, intimidating voters, en route to sweeping the election. The following September, Roncalli traveled to Bergamo on the tenth anniversary of Radini-Tedeschi's death—on the occasion of his remains being moved from a city cemetery to the crypt at the Bergamo Cathedral. The sermon he made that day was, according to one biographer, "the finest or the most foolhardy" he would ever make.

Speaking of patriotism and national pride, Roncalli said that such feelings were good and important for a nation, but that the

Church must determine "the true good of a country" in other ways, not just via "military enterprises, diplomatic agreements or economic successes," but by "justice embodied in law." Roncalli noted that differences in opinion between clergy and civil authorities should not preclude their cooperating, but he criticized Mussolini's concessions regarding religious education and crucifixes in public buildings as a kind of window dressing, particularly regarding education, since the state still had firm control of education beyond primary schools. He went on to say that, in any event, such changes were more due to the "champions of the just cause of educational freedom in Italy such as Radini-Tedeschi" than the Fascists.

Did Roncalli know that he was playing with fire? If so, to what extent? It seems that Roncalli was speaking more out of loyalty to the memory of Radini-Tedeschi than with any desire to tweak the Vatican or Mussolini, but his very real and instinctive distaste of the Fascists, whom Pope Pius was hoping to appease, comes through clearly.

Nothing happened in the immediate aftermath of the Bergamo sermon. The pope had made him a leading member of a commission to prepare for the coming Holy Year of 1925. Roncalli was put in charge of organizing lodging for thousands of pilgrims to Rome, a significant assignment that tested his organizational and executive skills. He was also asked to lecture on patristics, the study of Church fathers, at the Lateran Seminary, where he apparently shocked students by claiming, "In certain cases, it may be quite all right to sanction a mixed marriage [between Catholic and non-Catholic]."

But then, on February 17, 1925, the boom fell. Cardinal Gasparri told Roncalli the pope had named him apostolic visitor to Bulgaria, a predominantly Greek Orthodox nation with a small

Catholic minority. After a certain period of time there, Gasparri told him, Roncalli would enter the Vatican diplomatic corps and then be reassigned to a different and presumably more desirable—and more Catholic—country.

Roncalli immediately objected, saying that he was not a diplomat and that he did not have any experience in Eastern Europe, but Gasparri dismissed his objections away, as he did Roncalli's somewhat desperate suggestion that he could not go because it would leave his two sisters unemployed.

In a personal interview with Pope Pius XI on February 22, Roncalli learned the even more astonishing news that he would be made a bishop because, as the pope said, "it is not a good thing when an apostolic prelate goes to a country and has to deal with bishops without being one himself." In fact, as the pope's envoy, Roncalli would have the rank of archbishop.

All of this was a great deal for Roncalli to take in. While the new position was a promotion, it required him to leave Italy, in the same way that his mentor, Giacomo Radini-Tedeschi, was dispatched from Rome in 1905. It appears that he, too, had been tagged as a "modernist." (In fact, years later, asking to see his personnel file as pontiff, Roncalli found that it contained the notation, "Suspected of Modernism," which angered him so much he asked for a pen and then wrote, "I, John XXIII, Pope, declare that I was never a Modernist!" Later, cooling down a little, he had the humor to say, "I am the living example that a priest who has been placed under observation by the Holy Office can still become pope.")

Whatever the Vatican's motive in the matter, Roncalli had little choice. After installing his beloved sisters back home in Sotto il Monte (Monsignor Bugarini had died the previous year), Roncalli was consecrated archbishop on March 19, 1925. In the days leading up to the ceremony—held in Milan, at a church dedicated to Saint Charles Borromeo, Roncalli's favorite—he wrote in his journal, "I have not sought or desired this new ministry: the Lord has chosen it for me, making it clear that it is his will and that it would be a grave sin for me to refuse."

As with other great moments in Roncalli's life, he had not sought this change; it had sought him. And he would make the best of it. On April 23, 1925, he boarded the Simplon Orient Express in Milan and set off for his new assignment.

CHAPTER FIVE

Bulgaria, Greece, and Turkey
(1925–45)

When the newly consecrated Archbishop Roncalli arrived in the Bulgarian capital city of Sofia, he entered a country that was essentially in a state of war. During World War I, Bulgaria had sided with the Central powers. Because of their alliance, Bulgaria lost its territories in Macedonia, which included economically crucial access to the Aegean Sea. In the political turmoil that followed, communists and right-wingers struggled violently for control of the government. In 1924, there were more than 200 political assassinations.

A few weeks before Roncalli came on the scene, King Boris III was nearly killed in an assassination attempt. Though he survived, his prime minister was killed the next day. On April 17, at the prime minister's funeral in the Orthodox cathedral of Svate Nedelja, radicals set off a bomb planted high in the cathedral's dome, killing 150 mourners and wounding hundreds more. In response, Boris ordered his secret police to imprison, torture, and kill thousands.

As Roncalli stepped off the Orient Express, the local Catholic newspaper recounted the aftermath of the bombing: "Smoke lay over the entire country and nobody could see the way out. The blood of the victims and the tears of the afflicted are still fresh." Roncalli's first official act was to visit the victims of the bombing, regardless of their religion. This was much appreciated by those who watched the portly new envoy in action, but there was still a great deal of suspicion of the Roman Catholic visitor who was, after all, the first envoy the Vatican had sent to Bulgaria in 600 years.

The national census prior to World War II showed there were some 45,000 Roman Catholics in Bulgaria, plus 5,000 Uniates, followers of the Byzantine or Eastern Catholic, as opposed to the Latin, Rite. The largest religion, according to the same census, was Islam, with 780,000 practicing Muslims of either Turkish or Bulgarian descent. There were about 48,000 Jews in Bulgaria.

The Holy Synod, the ruling body of the Bulgarian Orthodox Church, complained through its newspaper that Roncalli's visit smacked of "imperialism" and might be the beginning of a plot to overthrow the Orthodox religion in the country.

Following his own instincts rather than instructions from Rome, Roncalli sought to dispel this suspicion. He first stayed with Monsignor Stefan Kurtef, a member of the Uniate group of the Eastern Catholic Church, which acknowledged the authority of the pope in Rome but used Byzantine liturgy. It was not quite like lodging with an Orthodox priest, but it was a step in the right direction. He also sought an immediate audience with King Boris, attempting to mend a relationship that had become badly frayed.

In the late nineteenth century, the Bulgarian royal family broke with the Church, when Boris's father, Czar Ferdinand, told Pope Leo XIII he was going to baptize his son Orthodox, the majority Christian church. According to various reports, an infuriated Pope Leo rose from his throne and pointed at the door, essentially kicking Ferdinand out of the Vatican.

The young Boris, at least, seemed open to talking to Roncalli, a minor victory, though a huge setback loomed. In the meantime, however, Roncalli set about his mission of learning more about Bulgaria's Catholics by journeying all over the country with Father Kurtef as an interpreter. It was in many ways an adventure.

The rough and mountainous back roads were impassable to motor vehicles, so Roncalli and Kurtef traveled by horseback, mule, and crude carts, spending the entire spring and summer traversing the country from the Black Sea to Turkey. It was a time he would always remember. The countryside was beautiful and the people welcomed him warmly, despite the fact that they were plagued by bandits, terrorist massacres, and extreme poverty. Rural Catholics were surprised to see him; no one from the Church except their lowly parish priests had paid any attention to them. They called Roncalli *Diado,* which meant "the good father," although sometimes they referred to him as "the round one," because his struggle with his weight had become decidedly pronounced. But he felt at home with Bulgarian peasants, who reminded him of his own family and the poor mountain people of Sotto il Monte.

Roncalli's sense of his mission in Bulgaria was twofold. First, he ministered to the needs of Bulgaria's Catholics, who were often widely separated in distance and who were in large part tended to by imperious French missionary nuns and priests. Roncalli ordered that prayers after Mass be in Bulgarian, not French, and

tried to put an end to rivalries between the numerous French religious orders, which included the Assumptionists, Vincentians,
Christian Brothers, and Capuchins.

The second part of his mission was to establish rapprochement
with the Bulgarian Orthodox Church; this was far easier said
than done. The schism between Eastern and Latin Churches had
lasted 1,000 years and remained wide. The Orthodox Church did
not recognize the supremacy of the pope. The Roman Church
recognized the validity of Orthodox sacraments but consistently
urged reconciliation (or "return") on the Orthodox faithful, an
attitude Orthodox Christians found patronizing and unrealistic.
Already bearing within his heart the seeds of ecumenism that
would blossom in the Second Vatican Council, Roncalli understood that he needed to approach the Orthodox faithful in a spirit
of love and respect rather than condemnation. He met with Patriarch Basil III in Sofia. He showed up unannounced at the Orthodox monastery at Rila, a city outside of Sofia, where he prayed
at its altar.

This was not the Vatican stance. On January 6, 1928, Pope Pius
XI issued his encyclical *Mortalium animos* (The Minds of Men)
on religious unity, which spoke of religions that "would willingly
treat with the Church of Rome, but on equal terms, that is as
equals with an equal: but . . . it does not seem open to doubt that
any pact into which they might enter would not compel them to
turn from those opinions which are still the reason why they err
and stray from the one fold of Christ."

The Church of Rome, in other words, would not bend to establish true parity with other faiths; despite this, Roncalli continued
to deal with those Orthodox faithful with whom he established
a relationship in Bulgaria via "the miracle of love" and "the pri-
macy of charity." And Roncalli's ecumenism was not at all of

the philosophical sort, either. In the spring of 1928, central Bulgaria suffered a series of severe earthquakes. Roncalli rushed to the region the very next day, assessed the situation—thousands homeless, severe flooding—and cabled the Vatican for emergency aid. He was in the city of Philippopolis (now Plovdiv) when another earthquake hit, and he spent the night on the street along with thousands of others.

In the next month, as he traveled back and forth to Sofia, he used money the Vatican had sent him, as well as any funds he could beg from friends back in Italy, to establish what became known as the "Pope's Soup Kitchens," which fed people in the devastated region throughout the next few months. Roncalli slept in tents among refugees. His empathy was as evident as it was profound. "Unfortunately," he wrote to his sisters Ancilla and Maria, "the continual rain has made life dreadfully hard for the numberless poor people who do not yet dare re-enter their tottering houses."

Roncalli's work with the victims of the earthquake brought tangible results, but he was not so sure about the rest of his efforts in Bulgaria. He successfully recommended that Monsignor Stefan Kurtef be made bishop of the Uniates—the Eastern Rite Catholics—so that a Bulgarian, rather than an emissary from Rome, represented them. But his larger plan to get the Vatican to approve a seminary to train Bulgarian priests failed. By 1930, five years after he arrived in Bulgaria, Roncalli had started to question his purpose there.

In his annual retreat in the spring of 1930, he wrote in his journal, "Make me love thy Cross," a thirteenth-century hymn to Mary.

A whole series of recent events has conferred on this retreat a special sense of loving abandonment to God, suffered and

crucified, my Master and King. The trials . . . have been
many: anxieties concerning the arrangements for founding
the Bulgarian seminary; the uncertainty which has now
lasted for more than five years about the exact scope of my
mission in this country; [and] my frustrations and disap-
pointments at not being able to do more.

Roncalli tried to reconcile the frustrations of his job, but it was
not easy. Most of the time, the Vatican paid scant attention to
him—he could not even get a reply about receiving extra funds
to enlarge the small house he lived in so that his sisters could join
him. When he did receive notice, it could be quite unpleasant. In
1930, King Boris decided to marry Princess Giovanna, the daughter of King Victor Emmanuel of Italy. The king approved of this
match, as did Mussolini, who thought it might help extend his
power into the Balkans, but Pope Pius balked because Boris was
Orthodox and Giovanna Catholic.

Roncalli was entrusted with the delicate mission of convincing
King Boris that he should accept the pope's conditions that the
marriage must be solemnized *only* in the Catholic Church and
that the children must be raised Catholic. The king was apparently willing enough, but the Bulgarian constitution specifically
stated that only an Orthodox prince could reign in Bulgaria.

Through careful negotiations spearheaded by Roncalli, Boris
eventually agreed to marry in a Catholic Church and raise his
future children in the Roman Catholic faith.

The Catholic ceremony took place on October 25, 1930. Six
days later, however, Boris held what he assured Roncalli was
merely an Orthodox "blessing" of the couple. In fact, it was an official Orthodox wedding ceremony, held in an Eastern cathedral
in Sofia. When he heard about this, Pius summoned Roncalli to

Rome, where he forced Roncalli to kneel before him while he berated him for embarrassing the Church. Roncalli later claimed he had informed the Vatican that they should not trust Boris under any circumstance.

This was not to be the last of Boris's apostasy. On January 13, 1933, Queen Giovanna gave birth to a daughter. With great pomp, Boris baptized her in the Orthodox Church. Though Roncalli lodged a formal protest, the king refused to receive him.

Roncalli exchanged heated words with the Italian minister to Bulgaria who had helped arrange the marriage. "Don't be so upset, Your Excellency," the minister reportedly told Roncalli. "After all, King Boris is Orthodox, and it's a matter for his conscience. As to Queen Giovanna, she'll just have to go to confession."

The queen, in fact, was more in need of Roncalli's consolation than his absolution. King Boris had taken the baby from her arms and had baptized her without the queen's consent.

In December 1934, Roncalli was named apostolic delegate to Turkey and Greece. He was ready to move on. Although he felt he had accomplished as much as he could in Bulgaria, he did not want to leave behind his many new friends. In his farewell Christmas sermon, delivered in Sofia, he made reference to the Irish custom of leaving a lit candle in the window to show the way to Mary and Joseph:

> Wherever I may go, if a Bulgarian passes by my door, whether it's night-time or whether he's poor, he will find that candle lighted in my window. Knock, knock. You won't be asked whether you're a Catholic or not. . . . Two fraternal arms will welcome you and the warm heart of a friend will make it a feast-day.

* * *

Roncalli arrived in Istanbul—or Constantinople, as many Christians continued to call it—in January 1935. There he faced at least as many challenges as he had when he had first set foot in Bulgaria. Of Turkey's 18 million people, approximately 79,000 were Jewish, and 100,000 were Orthodox Christian. Only 35,000 or so were Roman Catholic. The vast majority of the country worshiped Allah.

To complicate matters, the government of President Mustafa Kemal Atatürk, heirs of the Young Turks who overthrew the caliphate of Turkey before World War I, was in the middle of westernizing, or modernizing, the state. Now known as "father of the Turks"—a name he bestowed upon himself—Atatürk immediately banned religious displays of any kind, be they Islamic, Orthodox, or Catholic. He outlawed fezes, the distinctive, flat-topped and tassled hats worn by Muslim men, and, shortly after Roncalli arrived, he prohibited Turks from wearing religious habits.

Some apostolic delegates might have protested, but Roncalli knew when to pick his battles. "What does it matter," he wrote to a friend, "whether we wear the soutane or trousers as long as we proclaim the word of God?" A tailor made him a few dark business suits, and he started to wear a bowler hat. As one biographer wrote, he looked "like a Milanese bank clerk at the wedding of the managing director's daughter." He sent a few photos of himself in secular garb to his parents at Sotto il Monte with the note, "You will recognize your son the bishop dressed as the new law requires."

Roncalli understood his position in Turkey. He sent out no official letter announcing his arrival, telling his aides instead, "Let it be clearly understood here and now that in this country the apostolic delegate is a representative with no diplomatic standing." Roncalli knew he could do little about the Atatürk government's campaign against religion. He had to watch as Catholic schools were shut down one by one, replaced by state-run primary schools, but he could still continue the pastoral work that he loved—baptisms, celebrating Mass, delivering homilies—in different parts of the country.

Roncalli's work was interrupted in July 1935 by the very sad news that his father, Giovanni, had died at the age of eighty. Word came by telegram, and it caused Roncalli to go to his private chapel and "weep like a child." Unfortunately, he was unable to leave Turkey for his father's funeral, although he did return in September to spend time with his mother.

Roncalli's spirit of ecumenism served him greatly in Turkey. Conflicting ideas about religion and the state had brought about Atatürk's reforms, and antireligious fervor remained within the political sphere. Censorship of the press impinged directly on the archbishop's ability to communicate freely with Catholics. Roncalli expressed his hopes that he would still be able to preach about the need for charity. From the pulpit of the Cathedral of the Holy Spirit, Roncalli sought fit to preach in the language of his listeners. Officiating in Turkish at the Sacrifice of the Mass, an unprecedented act, reflected his desire to apply the ecumenical outreach in which he so fervently believed.

In 1936, he added the Turkish words *Tanre Mubarek olsun* ("Blessed be God") to the Divine Praises in the Mass. He also read the Gospel and Litany in Turkish. He saw this as a genuinely "catholic" step, but people left his services infuriated, and some even complained to Rome. "The difference," he wrote in his journal, "between my way of seeing situations on the spot and certain ways of judging the same things in Rome hurts me considerably; it is my only real cross."

Yet another cross, however, was his health. Though generally in good shape for a fifty-five-year-old, within his first year in Turkey he developed a small hernia that he left untreated. His weight also ballooned. "For my health's sake," he wrote in his journal in October, "I must stick to a diet as regards food. I eat little in the evenings already but now I must eat less at midday too. It will be good for me to go out for a walk every day. O Lord, I find this hard and it seems such a waste of time, but still it is necessary and everybody insists I should do so."

These same "everybodies" insisted he try yogurt for his health. "You see how it is," he wrote to his family. "We must become like children again and eat the food of the poor."

Even with all his concerns in Turkey, Roncalli did not forget that he was also apostolic delegate to Greece.

Greece presented distinctly different challenges. Roman Catholicism was not held in high regard there. Greece was Orthodox and considered Rome—and its insistence on Latin language and Latin traditions—an affront to its history and culture. Frankish hordes had brought the Crusades to the Greek peninsula in the thirteenth century; rape and pillage is not easily erased from the national psyche. In the twentieth century, political turmoil led to a military coup by General Ioannis Metaxas, who quickly became embroiled in the continued civil unrest on the streets and the vagaries of the various national agendas at play in the region. On October 2, 1935, Italy invaded Abyssinia (modern-day Ethiopia). Greece criticized the Italian invasion at the League of Nations and opposed the country's territorial ambitions in Albania, which infuriated Mussolini, straining relations between the two countries. Because of his Italian heritage, Roncalli was viewed with suspicion, even though he was an apostolic delegate from the Vatican.

Wherever Roncalli went in Greece, government agents followed. As in Turkey, he was careful not to overreach. He kept his visits brief, traveled inconspicuously, and never presumed he was more important than the people to whom he was speaking. There was a good deal of distrust for the Italians during this period. Greece was overwhelmingly Orthodox. For all intents and purposes, Orthodox Christianity was the state religion. There were only about 50,000 Roman Catholics, most of whom were viewed with suspicion.

Roncalli visited Catholics in Greece on numerous occasions. Though he traveled there three times in his first year in Turkey, he had a difficult time connecting with the Greek government, which did not recognize the Vatican envoy. Believing Mussolini and Pope Pius were allied, in part because of the Lateran Pact, General Metaxas refused to allow Roman Catholics to marry or to build churches. "There are so many things to fix in this country," Roncalli wrote to his mother in 1938, "but since the people are all Orthodox and frightened of the Holy See and the pope, one has to act slowly, cautiously, and with extreme sensitivity." During his time in Greece, Roncalli gradually, patiently, worked his magic over the authorities. He convinced the government to allow Greek Catholics to marry in their Church. They also allowed Uniate Catholics to build a new cathedral. As he did in Bulgaria, he continued to visit Greek Orthodox shrines, churches, and monasteries, even making the arduous journey to the famous monastery on Mount Athos. "One has to take the most difficult paths on horseback," he wrote to his mother, "and this made me entrust myself to Saint Joseph and my ancestors—as I always do—so as not to fall off. I didn't fall off once."

Outside Turkey, cascading events created a growing momentum in the direction of worldwide conflict in 1938. The ailing

Pope Pius XI issued two encyclical letters in March, denouncing the barren wastelands of Nazi idolatry and Soviet atheism. Roncalli reacted to the growing alarm as the dedicated pastoral leader he had always been and would continue to be. He was adamant in his belief that the Church must continue to seek understanding with the broader community of believers. Adapting to the needs of the locale and the requirements of the indigenous peoples was the key to the future.

Meanwhile, Hitler and Mussolini cavorted in Rome that May and appeasement reared its head in the signing of the Munich agreement in September. Then Kemal Atatürk died in Istanbul on November 10, 1938, a date commemorated still throughout the nation he forged as a modern state.

Nor did Roncalli "fall off his horse" on the mission to both Greece and Turkey, but in some ways there was very little headway an apostolic delegate could make in either country. As the world headed toward another great war, with Italy one of the aggressors, suspicions fell harder on the Church of Rome and its powerful prelates. Fewer and fewer doors were open to Roncalli. This was especially true after Mussolini's Ethiopian adventure. Italy and Greece traded belligerent words, and after Mussolini allied himself with Hitler, war broke out between the two countries in fall 1940.

Before that, however, Roncalli received a severe blow. His mother, Marianna, died of the flu on February 20, 1939, ten days after the death of Pope Pius XI. Roncalli mourned the pope, although Pius was a man he perhaps admired more than he loved. Marianna's death, however, crushed him. During her illness, he was unable to leave Istanbul and, even after her death, his duties did not allow him to return for the funeral. His sisters Ancilla and Maria wrote him, "Frequently she would recall her children, and especially you who are so far away; and you should have seen how, poor thing, she came to life again when she got a letter from you."

Marianna was eighty-five years old and had lived a full life, and Archbishop Roncalli consoled himself with this thought. In lieu of his presence at the funeral, he wrote the prose poem that appeared on the back of the funeral card:

> *Dear and respected by all.*
> *Dearer to her children*
> *Who grew numerous and strong*
> *In the fear of God and the love of men,*
> *And to the sons of her sons*
> *Whom she saw multiplied in joy*
> *In her home*
> *Even to the third and fourth generation*
> *Blessed her memory.*

Roncalli was by now the main support of his ever-growing family; often their needs exceeded what he could provide. "Without having taken a vow of poverty," he wrote ruefully to a friend, "I am practicing it." Years earlier, he purchased the family home and was still trying to pay off the mortgage and the yearly taxes. At the same time, he was forced to borrow money (12,000 lire in 1939 alone) to help support his sisters. He could occasionally sound a little irritable at the demands on his depleted purse. "But now for a little time I want to be left in peace and not appealed to for further needs," he wrote Ancilla, but it was evident how extraordinarily important his family was to him and how responsible he felt toward them.

In March 1939, Cardinal Eugenio Pacelli, the papal secretary of state, was elected pope as the obvious choice of the majority of cardinal electors after a one-day affair with only three ballots, the shortest conclave since the seventeenth century. He was the first Vatican secretary of state to be elected since Clement IX, 272 years earlier. "Being pope today," Roncalli remarked with typical humor, "is enough to turn your hair as white as your soutane." Pacelli took the name Pius XII.

Unable to be in Rome, Roncalli listened to the ceremony at Saint Peter's on the radio: "What a miracle the invention of the wireless is!" He had reason to be happy. Through Roncalli's efforts, a representative of the Ecumenical Patriarch Benjamin I was present at requiem for Pius XI—the first time the head of the Orthodox Church had attended a funeral Mass at the Vatican— and for the celebratory Te Deum for the new pope. As Roncalli wrote to his friend, the Anglican minister Austin Oakely, personal representative of the Archbishop of Canterbury, whose relationship with Roncalli was yet another example of the latter's spirit of ecumenism, there was a high wall between the Eastern and Western churches, but "I try to pull out a brick here and there." Once again, Roncalli's modest approach bore fruit.

On Good Friday 1939, Italy invaded Albania, exacerbating the growing tension on the Greek peninsula. On August 23, the Soviet Union signed a non-aggression pact with Nazi Germany, and on September 1, Germany invaded Poland, followed closely behind by their new allies, the Russians. World War II had begun, and Roncalli's world changed dramatically.

Soon England and France were at war with Germany; Mussolini, in a highly unpopular decision within Italy, invaded an already-defeated France in June 1940. When Mussolini attacked Greece from Albania in October, the Near East erupted. The outnumbered Greeks drove the Italians back into Albania, but at this point Germany and Bulgaria—recently brought into the war through King Boris's alliance with the Axis powers—invaded Yugoslavia and Greece. Both countries quickly surrendered despite the fact that Britain had joined the Greeks in attempting to fend off the Axis attack.

Roncalli, in neutral Turkey, watched these events unfold in horror. The Vatican ordered him to concentrate as much as possible on Greece, and he made three trips there in 1940—all now by plane, a cross for a man who hated flying. At one point, he arrived in Greece in June and did not leave until October. Despite the fact that he had reached the age of sixty, "the year in which a man begins to get old, and admits it," as he wrote in his journal, he was indefatigable. He visited Italian occupation troops, wounded German soldiers, and British prisoners of war and continually worked to provide relief for the hard-hit Greeks.

During the Axis occupation, 400,000 Greeks were imprisoned, 60,000 were executed, and millions were made homeless. The country was, as Roncalli wrote his family, "a place of desolation." The most pressing issue was food—the British blockade of Greece made it impossible for shipments of urgently needed grain to reach the island. Using Vatican funds, Roncalli established food depots, set up clinics for the sick, and begged for food and medicine from neutral countries.

But by September 1941, people were dying of starvation at the rate of 1,000 per day. Roncalli was asked by a delegation of Greek laymen if he could get the Vatican to intercede with the British to allow food shipments through the blockade. He immediately went into action, contacting Archbishop Damaskinos Papandreou, the only effective Greek leader in a country that had seen its king and his entire government flee to London.

The archbishop passed Roncalli a letter, signed by Orthodox leaders, that begged Pope Pius to use his influence to help Greece.

Roncalli personally delivered this letter to the pope at a private audience in October, and the Holy Father agreed to press Britain to allow the grain through to the starving Greeks.

Historians of the war disagree over the outcome of Roncalli's and Pius's intervention—some claim that the grain helped avert a worse famine, others that only a trickle got through. But what is indisputable is that Roncalli's pressuring of the Vatican at least alerted the Holy See to the scale of the humanitarian crisis in Greece, causing it to funnel more funds to that country.

It was at this point in the developing catastrophe that Roncalli began to expand his charitable ministry. The apostolic delegate was dedicated to performing corporal works of mercy among the people. Obedience and charity were the cornerstones of his vocation. Negotiating diplomatic terms was one thing, providing for the basic necessities of life was entirely another. It was the gospel mandate by which he lived.

Diplomacy, however, remained a critical element of his office. Dictated by the neutrality of Vatican policy, his methodology nonetheless was geared to a more personal interaction between him and the important players with whom he came in contact in Istanbul and Athens in these early months of the war.

One of these players was Franz von Papen, German ambassador to Turkey. Von Papen was a holdover from the Weimar Republic. He had been one of a group of politicians who had supported the Nazi assumption of power in 1933 in the hope that Germany could benefit from the discipline espoused by National Socialism. Deputized by Hitler to oversee the annexation of Austria in 1938, he was now in Istanbul at the crossroads of East and West. The relationship that developed between the Catholic von Papen and Archbishop Roncalli illustrates the ambivalent nature of the diplomatic process.

Von Papen would serve as a conduit, both for information and misinformation, as well as a link in the escape route for Eastern European Jews. Throughout 1940, Roncalli traveled between Istanbul and Athens attending to the many exigencies the two locales presented. Germany's territorial ambitions in Western Europe were actualized on June 22, 1940, when the French signed an armistice at Compiègne. Earlier that month, Italy officially entered the war as the third belligerent of the Axis powers.

In February 1942, Roncalli wrote, "I live in the exercise of charity, charity for all." As the situation worsened in Greece, the apostolic delegate spent most of his time in Istanbul, where he worked tirelessly to save lives. He begged the German commander in Athens to spare the lives of Greek partisans captured there, but he failed to sway the Nazis, and the executions went on as ordered. Roncalli also spent much of his time working with the Red Cross to track prisoners of war and provide their families with information.

As the war progressed, Istanbul was a hotbed of espionage and a way station for those attempting to escape the onslaught of Nazi aggression.

During the early years of the war, Roncalli became aware of Germany's campaign of extermination of European Jews, probably from reports by Polish refugees. He repeatedly forwarded requests to the Vatican from Chaim Barlas, head of the Jerusalem Jewish Agency, to pressure neutral countries like Portugal and Sweden into accepting displaced Jews.

Barlas also wanted the Vatican to aid in "the transfer of Jews to Palestine."

Secretary of State Cardinal Maglione stated confidentially, "One cannot prescind from the strict connection between [Jews reaching Palestine] and that of the Holy Places, for whose liberty the Holy See is deeply concerned." In other words, Pope Pius seemed to be more interested in preserving Jerusalem and its sacred places for Christian pilgrims than in saving Jewish lives.

Despite this indifference—to put the very best light on it—on the part of the Vatican diplomatic machinery, Roncalli did everything he could.

As the war entered its second year, severe shortages of food in Greece became an overriding concern for the archbishop. The Allies were enforcing a total blockade of supplies to belligerent countries in pursuit of unconditional surrender by the Axis powers. Allied war aims were not part of Roncalli's mission; feeding hungry people was. To alleviate the pressing need for food stores in Greece, he met with the Orthodox Metropolitan Damaskinos, a meeting not previously considered, such was the historical hostility between the two rival churches.

Roncalli's involvement with relief efforts introduced him to Raymond Courvoisier, coordinator of the Red Cross/Red Crescent efforts in Ankara. Attempting to mediate on the part of prisoners of war held in the Soviet Union, the archbishop became acutely aware of the hostility inherent in Soviet policy toward humanitarian initiatives. The position of the Vatican as protector of the suffering was not held in high regard within the Soviet foreign ministry, which focused almost singularly on Soviet power.

Roncalli's relationship with Courvoisier, however, became crucial to his efforts to save European Jews. Turkey remained a neutral country, and Istanbul a critical point of departure from blood-soaked Europe. Approached by members of the Jewish Agency from Palestine, Archbishop Roncalli met with the Grand Rabbi of Jerusalem, Isaac Herzog. Formal requests of the Vatican to intercede with Great Britain on the issue of the Mandate for Palestine and the concomitant restriction on immigration were unsuccessful. It became incumbent upon Roncalli to facilitate action of an independent nature. And so he did. It is a matter of record that Archbishop Roncalli personally signed transit visas for thousands of Slovakian Jews, whose destination and eventual refuge was, indeed, Palestine.

Assisting Archbishop Roncalli in this essential work of mercy was von Papen, Hitler's emissary in Turkey. The personal relationship between the two men enabled Roncalli to effectively channel resources to German-occupied areas.

Despite his alliance, King Boris responded to Roncalli's personal pleas and delayed the transfer of thousands of Jews to concentration camps. He arranged for transit visas to Palestine. As many as 24,000 Jewish lives were saved in this way. Unfortunately, Boris died mysteriously after a visit to Hitler in Germany—he may have been poisoned by the Nazis for not being loyal enough—and Roncalli was unable to save additional lives in the Balkans. In 1944, Roncalli forwarded to Vatican diplomats in Hungary and Romania so-called immigration certificates prepared by the Palestine Jewish Agency. According to legend, these were forged baptismal certificates prepared by Roncalli himself. This was not the case. Roncalli was not in the business of faking baptisms, even to save Jewish lives. But he did aid in getting the immigration certificates to Jews in these beleaguered countries, where such paperwork sometimes helped them escape.

Maintaining relationships within German-occupied countries and retaining diplomatic status in Axis-controlled capital cities were fundamental goals of the Vatican diplomat throughout the duration of the war.

Pius XII had served as nuncio to Berlin and as secretary of state immediately prior to his election as pope. Cardinal Luigi Maglione, Pius's secretary of state during World War II, priori-

tized Vatican diplomatic outreach and considered the plight of
Jewish refugees as secondary to the position of the Holy See as
an international diplomatic institution. Critical historians have
posited that political influence outweighed considerations of humanitarian relief.

In December 1944, Roncalli's life then took another fateful
turn. He received a cable from the Vatican that read, "His Holiness nominates you nuncio to Paris. Letter follows."

At first, Roncalli thought the cable was a mistake, but his secretary confirmed its authenticity. Here was the invitation to serve as
an official ambassador, the Vatican's top diplomatic post to France,
a country whose future was very important to the Vatican. He did
not know that Pope Pius's first choice had turned him down for
health reasons. He just knew that, at the age of sixty-three, he was
on his way to the most important job of his career.

Chapter Six

France and Venice (1945–58)

After saying his farewells in Istanbul in late December, Roncalli flew first to Rome, where he met with Monsignor Domenico Tardini, the acting secretary of the Office of Extraordinary Affairs of the Holy See, the foreign department of the Secretariat of State. Roncalli knew Tardini well. He was a compact, bullet-headed man with tremendous energy, ability, a warm heart, and, according to one biographer, a "caustic sense of humor."

"Look here, old friend, are you sure that cable I received is not a mistake?" Roncalli asked. "Surely the Holy Father did not intend to appoint me nuncio to France."

"It's no mistake," Tardini replied. "His Holiness makes his own decisions. But you can be sure of one thing: None of us thought he would do this.

"You're not the only one who's astonished." He pointed toward the papal apartments, where Roncalli would soon meet with Pius. "It was all *his* idea."

Not exactly reassuring. Pius was warmly gracious to Roncalli when they met on the morning of December 29, making a point of telling him, "I want to make it clear that I was the one who acted in this nomination, thought of it and arranged it all. For that reason you may be sure that the will of God could not be more manifest and encouraging."

The pontiff made it clear that he would rely on Roncalli's proven tact and judgment of relationships to help restore the position of the Church in France, though he did not expect the new nuncio to work miracles. "In your position," the pope said, "you can only do, as they say in France, *aussi bien que possible*."

Of course, as with all of Roncalli's appointments to higher episcopal offices, there were politics at play. In late 1944, France was only a few months removed from Nazi occupation.

Through most of World War II, France had been run by Marshal Philippe Pétain. His Vichy government, which owed its existence to the Nazis, actively collaborated with Germany. It formed its own secret police, known as the Milice, to help suppress the French Resistance, often with brutal results. Members of the force even rounded up Jews. At the same time, the Free French forces, based in England and North Africa prior to the invasion of Normandy, coordinated with the French Resistance to fight the Germans throughout France during the war. To the east, Allies continued to battle a resilient German Army in the Belgium ports.

After the liberation, General Charles de Gaulle, the charismatic leader of the Free French Army, became the provisional leader of France. One of his first acts as president was to imprison and, in some cases, execute former members of Pétain's Vichy regime who collaborated with the Nazis during occupation. The politics of liberation assumed preeminence in all matters of state for de Gaulle, which initiated a round of gamesmanship with Pius XII. A few months earlier, in August, the Vatican renominated Archbishop Valerio Valeri, nuncio to Pétain's regime, but the French leader wouldn't hear of it. Valeri was too closely tied to the discredited Vichy government, although he personally had not collaborated with the Germans. General de Gaulle, the most respected man in the country, maintained that liberated France needed someone new.

At first, Pius refused to budge, sticking to Valeri as his choice for nuncio. He and de Gaulle, two notably stubborn leaders, remained at an impasse throughout the fall and early winter. But then the pope learned that the Soviets had already recognized the de Gaulle government, sending their official ambassador to Paris. Since 1815, following the Congress of Vienna, the papal nuncio had been considered dean of the diplomatic corps, regardless of his length of service. With that honor came the right to offer New Year's greetings to the French head of state in an official ceremony.

If Pius did not confirm a nuncio by the end of the month, then the Soviet ambassador, as the next most senior official there, would present greetings to de Gaulle. More than anything on earth, the pope abhorred the danger presented by godless communism. It would simply not do to have the Soviets steal a march on the Vatican.

But when Pius's first choice as a substitute, Archbishop of Argentina Joseph Fietta, declined for reasons of poor health, he needed someone very quickly and picked Roncalli, who had performed well and gained a reputation for equanimity, despite serving in the backwaters of Bulgaria and Turkey and Greece. Roncalli also spoke French, and his relatively liberal leanings

would stand him in good stead in a country that was trending to
the left politically.

Still, as Roncalli boarded Charles de Gaulle's personal airplane
in Rome on the morning of December 30—for de Gaulle was as
anxious to get Roncalli to supersede the Soviet ambassador as the
pope was—he knew he wasn't anyone's first choice. As he would
write ruefully to a friend, "When the horses break down, they trot
out a donkey." And to the bishop of Bergamo, his home diocese,
he wrote:

> I seemed to be seized by surprise, like Habakuk, and trans-
> ported suddenly from Istanbul to Paris by a sort of incanta-
> tion. Also my interior discipline was turned topsy-turvy . . .
> the more so since it seemed absolutely incredible to me and
> certainly I had neither the courage nor the imagination nor
> the desire for it. I was stupefied.

Once in Paris, Roncalli began his whirlwind round of duties
on New Year's Day 1945, when he presented his credentials to
de Gaulle and in turn officially recognized the French leader's
new government on behalf of the international diplomatic corps,
lavishing compliments on the general and his staff and praising
their patriotism. "I think it went off very well," Roncalli noted
in correspondence with a friend, pointing to the evidence of de
Gaulle's gratitude and emotion, despite the president's impassive
outward expression.

It was typical of Roncalli that he made a point of apologizing
to the Soviet ambassador for upstaging him; he also made sure to
call on the Soviet embassy the very next day. When these diplomatic niceties were over, however, Roncalli got down to the tricky
business of serving as papal nuncio in Paris.

The first problem he encountered was the French government's desire to rid itself of "collaborationist" clergy. De Gaulle presented Roncalli with a list of twenty-five bishops who he said worked with the Nazis. He wanted them recalled to the Vatican. This list included Cardinal Emmanuel Suhard, archbishop of Paris, and numerous other prominent prelates. Examining the charges against them, Roncalli saw no real evidence, just "newspaper clippings and gossip," as he wrote to the French foreign minister. It was true that the bishops had told their flocks to cooperate with the Vichy government, but this had been the Vatican stance as well. And some of the bishops who had early on preached coexistence with the occupying Nazis had spoken vigorously against them as soon as they realized how perfidious Hitler's regime really was.

By the time Roncalli got involved, de Gaulle had realized that the removal of so many prelates without hard evidence was going to be impossible. So he told André Latreille, a Catholic historian and "director of cults" in the Interior Ministry, to find "four or five" bishops to expel as examples. Latreille went to visit Roncalli to discuss this matter in February and later painted an interesting picture of the new nuncio in his diary: "[He is] a very lively talker, stout, friendly, words tumble forth from him so that it is hard to get a word in edgeways."

Despite this cordiality, Roncalli was firm with Latreille. "We have to say exactly what we want, and we have to produce evidence. And we mustn't expect the new nuncio to become the Torquemada of French bishops."

To his Vatican superiors, Roncalli suggested that Jules Saliège, the archbishop of Toulouse and a favorite of Resistance forces, be made a cardinal—which he was in the next consistory in February 1946.

"My role in France is like that of Saint Joseph," he told Jacques Domaine, chief of protocol in the French Foreign Office, "to be guardian over our Lord and to protect him with discretion."

This is a perfect snapshot of Roncalli's style—friendly, engaging, compromising, but concerned with fairness. He refused to take part in a witch hunt of prelates. In July 1945, the Vatican quietly removed seven bishops from France, forcing them into retirement, though Cardinal Suhard remained in place. Roncalli was not solely responsible for this smaller number—de Gaulle saw the prudence of backing away from a wholesale removal of clergy—but the new nuncio certainly helped lessen the tension around the affair and toned down the heated rhetoric.

As nuncio, Roncalli lived, as he wrote his family, "in a princely palace with everything one might need, two secretaries, three nuns, three staff, five servants and a splendid car [a Cadillac]." Unlike in Bulgaria, Greece, and Turkey, his job was not to minister to the faithful, who, in any event, were well served in a predominately Catholic country. But it was not like him to spend all of his time in diplomatic meetings. As Monsignor Loris Capovilla, the man who would become his closest aide and confidant, later noted, "As nuncio, Monsignor Roncalli did not shut himself up in an imposing palace at the end of a street in Paris, but went every day in search of a community, or a soul, that for him was France."

Biographer Alden Hatch paints an intriguing picture:

His luncheons and dinners soon became famous in Paris, not only for the fare, but for the conversation. With the nuncio leading the talk with his deep knowledge of politics, art, and literature, and, above all, his spiritual inspiration, combined as it too rarely is with wit, the world gathered to feast its ears as well as its mouths. Regulars at Roncalli's table included many of the men who were shaping the destiny of France. Among them were Bidault, of course, and such other cabinet ministers—past, present, and future—as René Mayer, Edgar Faure, Robert Schuman, René Pléven, Antoine Pinay, men representing almost the whole kaleidoscope of French politics, as well as such great men of letters as François Mauriac. Perhaps Roncalli's greatest friend was the grand old socialist and anti-clerical, Eduard Herriot. The archbishop's greatest spiritual triumph, because heaven so rejoices in the return of lost lambs, was when, on his deathbed, this upright defender of secular philosophies asked for and received the sacraments of the Catholic Church.

Roncalli's favorite "infidel" was Naman Menemengioglu, the Turkish ambassador to France. It is difficult to determine the exact Muslim population of France, but it is clear that their number was very small. But that total differs if one considers French Algeria to be an integral part of France, as the postwar governments did until the nation's independence in 1962. In 1940 there were some 7.6 million people in Algeria, the vast majority of whom were Muslim. Unlike Jews, these Muslims were not persecuted under the Vichy regime, although they were subject to the apartheid-like laws already in effect under prior French rule.

Roncalli preached at parishes all over the country on feast days and special occasions and went out of his way to meet people. He journeyed throughout France, visiting each of the country's eighty-two dioceses, with the exception of two. In Paris, he walked whenever possible.

Once, according to an oft-told tale, he overheard a worker swearing vehemently in the nuncio's new apartment. Aides were horrified at the worker's inventive stream of profanity, but Roncalli walked up to the man and asked, "What is all this, my good man? Why can't you just say *merde* like everyone else and get on with your work?"

He also visited French North Africa, driving from Tunisia, through Algeria, then through Morocco to the Pillars of Hercules on the Mediterranean Sea. His French friends said that, like the Roman general Scipio, he should assume the title "Africanus."

One of the most influential French cardinals, Eugene Tisserant, prefect of the Sacred Congregation of the Oriental Church, became Roncalli's great friend. He lived in Rome but made many visits to Paris during the nuncio's term there. They conversed for long hours, with Tisserant sharing Roncalli's abiding ambition to pave the way for the eventual unification of the Orthodox churches with Rome. In the conclave to come, Tisserant's support, and that of the other French cardinals, would be critical to the success of the rotund Italian diplomat.

Roncalli's French was another subject for amusement. In general he spoke it fairly well—although some said with a bit of a guttural accent that made him sound like a Russian—but he knew he was not always on the mark. As he began a sermon in one Paris church, the microphone acted up, emitting loud electronic squeals, so he came down to the floor of the church and said, unamplified, "Dear children, you have heard nothing of what I was saying. That doesn't matter. It wasn't very interesting. I don't speak French very well. My saintly old mother, who was a peasant, didn't make me learn it early enough."

In fact, it was his old mentor, Bishop Radini-Tedeschi, who had taught him to speak French, but this type of humor and modesty on the part of Roncalli endeared him to most Frenchmen, as did the fact that he at least tried to speak their language.

Roncalli continued to be preoccupied with his aging. He wrote in his journal, "I must not disguise from myself the truth. I am definitely approaching old age. My mind resents this and almost rebels, for I feel so young, eager, agile and alert. But one look in the mirror disillusions me. I must do more and better, reflecting that perhaps the time still granted to me for living is brief."

Yet there is the strong sense that Roncalli enjoyed himself on a larger world stage and was growing as a diplomat and prelate. There are glimpses of him in numerous letters and memoirs holding dinner parties in his renovated residence on the Rue Président Wilson, bringing together French politicians, important clerics, foreign diplomats, and the like. People came for the excellent dinners but stayed for Roncalli himself. Robert Schuman, who later became French premier in 1947, commented on Roncalli, "He is the only man in Paris in whose company one feels the physical sensation of peace."

This sense of peace sometimes belied the turmoil Roncalli had to deal with in France. In some ways, he was leading the life of a nineteenth-century diplomat while the country changed drastically around him. Socialism, abhorred by Pius XII, was sweeping through Europe. One manifestation of it in France was the priest-workers movement. During the war, when the Nazis marched 800,000 young Frenchmen off to forced labor camps in Germany, Cardinal Suhard secretly assigned twenty-five young priests to go with them, disguised as workers, to minister to them. Most of these priests were discovered, and two were to die in concentration camps. Other priests spent time in concentration camps, working together with other captives of the Nazis. Based on their experiences, with barriers broken down between cleric and laypeople, they felt a truer communion—a more Christ-like connection—could be attained.

At the onset of the German occupation, there had been about 350,000 Jews in France. Many were not ethnically French but were refugees from Germany or other Nazi-controlled regions. While Pétain had quickly introduced anti-Jewish legislation, many ethnically French Jews managed to survive the war. Some historians argue that Vichy protected French Jews, deporting instead the foreigners first; others believe that Vichy realized that Hitler would eventually demand the deportation of all French Jews. About 273,000 of the original 350,000 survived, meaning that some 77,000 were murdered in the Holocaust.

At war's end, there was little to celebrate, other than survival. Some returning priests (mainly Parisians) sought jobs in factories, mingled freely with workers in cafés and restaurants, and in general refused to live lives circumscribed by the old formalities of the Church. Many French Catholics—including Cardinal Suhard—supported these men. Others were shocked, reporting to Roncalli that there were priests saying Mass in factory clothes and greeting their congregation with a jaunty, "Hi, guys!"

Roncalli himself practiced caution when replying to these complaints. He valued traditional Church life, which included the separation of a priest from his congregation. At the same time, however, one senses that his humanist heart understood what the priest-workers were about. They were not necessarily communists or socialists, though some did lean in this direction. At heart, they sought a truer essence of what it meant to follow the life of Christ, which echoed the spirit and concerns of Radini-Tedeschi's Catholic Action groups, the anchor of Roncalli's life and ministry.

Although as pope he would later come out more clearly on the side of the priest-workers—or at least, on the side of the *idea* behind their movement—Roncalli did not quite make his opinions known at this time. But once again, he was open to change within a framework of tradition. This openness, from a deeply pious and conventional priest devoted to doctrine, became the hallmark of the man—and the pontiff of Vatican II.

While nuncio, Archbishop Roncalli resumed his habit of vacationing in Sotto il Monte and continued his role as paterfamilias of the Roncalli clan. He sent Giuseppe, his youngest brother and now a father of ten, money to fix his teeth, since, as he wrote, "I have noticed that all we Roncalli brothers suffer with our teeth. . . . But with poor teeth [Roncalli had his pulled and replaced by dentures in his fifties] a man eats food badly and digests it even worse." And he followed up, too, writing Giuseppe's daughter Enrica to make sure her father put his teeth in every morning.

At the same time, he helped out a family cousin with a sick wife; refused to intercede in a minor court affair involving another cousin (although he did send him 20,000 francs); and brought his family a radio, connecting them to the world beyond the mountains.

Despite his protestations of age, intimations of mortality, and self-reproach in his journal (he had, he wrote, "to watch myself closely to overcome my natural sloth"), he was incredibly active. In 1950, having traveled through most of France, he set out on a 6,000-mile automobile trip through Spain and North Africa to visit French Algeria, which was torn by violence and civil war. Upon his return, Pope Pius issued his encyclical *Humani generis* (Of Mankind), which was critical of neo-modernist theologians who were exploring ideas not sanctioned by the Church. Many of these theologians taught in France.

The encyclical represented a kind of crackdown; these Catholic priests, teachers, and philosophers (who went unnamed, but everyone, including them, knew who they were) were drifting away from orthodoxy. When Church doctrine needed to be clarified, Pius said, it was clarified by encyclicals such as the *Humani generis,* which "pass judgment on a matter up to that time under dispute, [and then] it is obvious that the matter . . . cannot be any longer considered a question open to discussion."

The proponents of *la théologie nouvelle,* the "new theology"—who included Jesuits who admired the existential philosophy of Father Teilhard de Chardin, and others, mainly Dominicans, who believed in the importance of historicity (as much as divine intervention) in the lives of Church fathers—were hit hard by this. Many were forbidden to teach and were cycled to dead-end jobs within their religious orders. Pius XII's encyclical represented a conservative backlash akin to that of Pius XI. Interestingly, Roncalli was silent on this matter. Although, as pope, he would tolerate radical theologians ("Without a touch of holy madness," he later said, "the Church cannot grow"), he was much more interested in basic matters of faith—such as charity, tolerance, peace, and love of one's fellows—that touched the lives of everyday men and women.

Still, he did not like to see honest debate repressed, and, as pontiff, he would bring some of these same disgraced scholars back to play important roles in Vatican II.

Along with everything else he was doing, Roncalli became the official Vatican observer to the United Nations Educational, Scientific and Cultural Organization (UNESCO), which was headquartered in Paris—and which he conditionally embraced. Nonetheless, Roncalli was not blind nor naïve: from the first, he recognized the danger that UNESCO would tend to propagate a mass materialistic culture rather than encourage the precious individual spiritual and artistic values of its many nations. Addressing the general conference of the organization in July 1951, he told the delegates, "UNESCO is a great burning furnace, the sparks from which will everywhere kindle . . . widespread cooperation in the interests of justice, liberty and peace for all the peoples of the earth, without distinction of race, language or religion. . . ." The delegates were surprised and pleased that Roncalli did not take a more dogmatic Catholic view; his ecumenical views were already coming to the fore.

He expressed his thoughts in this maxim: "To look at each other without mistrust; to come close to each other without fear; to help each other without surrender."

Author Alden Hatch, in *A Man Named John,* again provides a snapshot of the colorful impression this unique practitioner of diplomacy made on the world stage among his international confreres:

> Soon his rotund figure in its swinging, tent-like cape became as familiar in the chilly halls and conference rooms on the Avenue Klèber as it was in the bookstalls on the quays or the salons of the Rue du Faubourge Saint Honoré. His ability to speak languages such as Bulgarian, Turkish, modern Greek and even a little Russian endeared him to the delegates from those countries. Indeed, there is a photograph of him in amiable conversation with his great opponent, Soviet Ambassador Bogomolov, showing that even there he sought "what unites rather than what divides."

In November 1952, Roncalli received a letter from Monsignor Giovanni Battista Montini, a good friend who served as sustituto, or acting secretary of state for Ordinary (that is, internal Church) Affairs within the Secretariat of State, the Holy Father's right hand. (Montini would later succeed Roncalli as pope, taking the name Paul VI.) His letter asked in great confidence if Roncalli would consider succeeding the current patriarch of Venice, Carlo Agostini, who was gravely ill. It would mean a cardinal's red biretta for Roncalli and a return to his beloved Italy. Roncalli was instructed to tell no one. He recorded in his journal, "I prayed, thought about it, and answered *Oboedienta et pax* [Obedience and peace]."

Then, later in November, he was informed that the pope would "elevate him to the sacred purple" (an archaic reference to the princely rank of cardinal) in January. Hearing that his sister Ancilla was mortally ill, he raced to her bedside in Sotto il Monte to comfort her, but she rallied and in fact would live nearly another year. On December 29, he opened the morning paper to learn that Patriarch Agostini had died at age sixty-four. "With this death," he wrote in his journal, "the new direction of my life . . . begins."

Invested as a cardinal at the consistory on January 12, the seventy-one-year-old Roncalli arrived in Venice to a tumultuous welcome on a sun-drenched March 15, 1953. For Roncalli, a lover of tradition and ceremony, it was heaven. The citizens turned out on the banks of the canals in huge numbers; Roncalli, wearing scarlet robes and a short ermine cape, rode in the lead vessel of a flotilla of gondolas and motor boats, waving and smiling at the cheering faces all around him. In his sermon at the ancient Saint Mark's Basilica, he told the assembled crowd, "I commend to your kindness someone who simply wants to be your brother."

Venice was a brilliant city with an extraordinary history. Settled by Romans fleeing barbarian invasions as the Roman Empire crumbled, it was a city floating on water, once a great seaport and bastion of art and music. But now many of its palaces were old and shoddy, and its young people were leaving to seek work elsewhere. It had become a scenic tourist destination—home to famous music, art, and movie festivals—but Roncalli discovered that among his flock of 400,000 there was widespread unemployment.

Shortly after becoming patriarch, Roncalli wrote in his journal, he discovered "two painful problems . . . amidst all the splendor of ecclesiastical state, and the veneration shown me as Cardinal and Patriarch: the scantiness of my revenue and the throng of poor folk with their requests for employment and financial help." At the same time, he remained astonished at how far he had risen: "The arc of my humble life, honored far beyond my deserts by the Holy See, rose in my native village and now curves over the domes and pinnacles of Saint Mark's."

Saying that he was "like the mother of a poor family who is entrusted with so many children," Roncalli set to work. He made his drafty living quarters as habitable as possible and found the people he needed to help him—nuns to look after his household; a young layman named Guido Gussi, who became his personal butler; and a thin young priest named Loris Capovilla, who became his most trusted aide and, later, his literary executor. He sold the patriarch's summer palace to pay for a new seminary and used personal funds to support charities for the poor.

As in Paris, he did not shut himself away in his magnificent marble residence. Instead, he spent a good deal of time in public, where he enjoyed traveling on the *vaporetto,* the water bus, rather than in the Fiat provided him by the diocese. Because the Vatican wanted to make sure that the Communist or Socialist political parties were beaten back by the Christian Democrats, he could not avoid getting involved in local politics.

Cardinal Roncalli's routine, such as it was, took him to every parish church within his patriarchate—most more than once—and on pilgrimages beyond its borders. Even at home in Venice, the religious ceremonies and duties were so numerous and varied that he had little time of his own.

His predecessor, Patriarch Carlo Agostino, had worked incredibly hard, driving himself by sheer willpower—indeed, pretty much driving himself to his own death. As a result, he had been irritable, giving off an impression of arrogance and lack of patience. Roncalli, with his powerful peasant constitution, always seemed unhurried and patient. He once said, "There's nothing wrong with my liver and nothing wrong with my nerves, so I enjoy meeting people." The Venetians enjoyed meeting him, too. They jokingly called him "the calm after the storm," and they saw a great deal of him.

Despite his surroundings in the episcopal palace of a near-prince—with its ornate halls and salons, the exquisite chapel with its *cinquecento* embellishments, his study with silk-covered walls, heavy carved furniture, and more than 400 books (which he delighted in showing to visitors)—he maintained the personal simplicity and modesty that had marked him from his days as a seminarian in a simple cell with a stark desk and a handful of textbooks.

In November 1953, to Roncalli's great sorrow, his sister Ancilla died. He went back to Sotto il Monte for the funeral, accompanied by his aide, Loris Capovilla. It was a chill autumn day, not unlike the autumn day he was born, with the *tramontano* blowing dead leaves through the simple village graveyard. His sister had spent much of her life in his service; before the coffin was closed, he kissed her on the forehead, remarking to Capovilla, "That's the second time I've done that—the first was when I knew she was dying." On the train ride back to Venice, Capovilla, sitting next to Roncalli, heard him mutter to himself, *"Guai a noi se fosse tutta un illusione."* (Woe to us if it's all an illusion.) It was an enigmatic comment: Was the patriarch of Venice wondering if God himself was a human construct? Roncalli may have been feeling some guilt over his sister's long life in his service, without children or family of her own, or he may have been once again feeling intimations of mortality.

Capovilla, who knew Roncalli quite well, wasn't sure what the patriarch meant. The comment revealed, he later wrote, "a disconcerting aspect of genuine humanity in my patriarch." If it was all an illusion, Roncalli had spent fifty years in its service, for the following August he celebrated a half-century as a priest.

His nephew Battista, a somewhat hapless young priest, organized a celebration for Roncalli back in Sotto il Monte, but when Roncalli heard from Capovilla that Battista planned something elaborate, he wrote him a furious letter: "I do not want, I do not desire that anything more should be done than in previous years," he told his nephew. "Why do you have to offend me and make me suffer? Are you trying to stop me from coming at all?"

It was an uncharacteristically ill-tempered letter from Roncalli, but it had been a tough year for him. Two months after Ancilla's death, his sister Teresa died suddenly, and another sister, Maria, contracted the cancer that would kill her in early 1955. Another reason that he did not want elaborate ceremonies is that he was aware that his name was being mentioned in certain circles as a possibility to become the next pope. As cool and diplomatic as he was, the patriarch always kept his ear to the ground on such matters. After all, he would likely be an elector himself in the next conclave.

Pius XII nearly died in 1954; although he recovered, the state of his health continued to be uncertain. Both flattered by and fearful of the idea that he might one day become the Roman pontiff, Roncalli was horrified that people might think he was putting on airs at his golden jubilee celebration.

However, concern over what people thought typically didn't get in the way of Roncalli's own political opinions. As the papal secretary, Monsignor Gino Spaento, later pointed out, "No one should be deceived by the patriarch's simplicity into thinking him a simpleton. His simple manner is the result of his holiness, but he is a very complex and profound personality, keen, alert and anything but stupid."

In matters of principle and Catholic doctrine, Roncalli was immovable. The Vatican required him to publicly condemn an "opening to the Left" on the part of the Christian Democratic Party in 1956 as "a most serious doctrinal error and a flagrant violation of Catholic discipline," but when a Socialist congress was held in Venice in early 1957, Roncalli went on record hoping that the Socialists would "do everything possible to improve living conditions and social well-being." It just was not in his nature to turn away those whom he felt might do some good; besides, he did have ulterior motives.

As he told a Venice newspaper at the time, "No one should be disturbed by my initiative. One day all those people [the Socialists] will come to church again." Though he abhorred the socialist philosophy, he loved Socialists as fellow men created in the image and likeness of God.

Perhaps the happiest time in his life was Roncalli's term as patriarch of this ancient See. In those years he was finally able to realize his long-held desire to be, above all, a priest and minister of souls. The administration of a diocese and the thousands of demands on his time and attention were, to him, secondary to the primary task of tending to the spiritual needs of the people, "into which he flung himself with all the energy and ardor of his nature. He was, in fact, joyfully fulfilling a spiritual need of his own."

Although Roncalli avoided the subject publicly, he knew that Pope Pius XII had never quite recovered from his illness in 1954. As the summer of 1958 drew to a close, the pontiff fell quite ill again. In October, Venice was host to a medical convention, and the Vatican contacted Roncalli for help in finding an international specialist in internal disorders who was attending the convention. Later, the Vatican Secretariat of State privately notified the College of Cardinals—there were fifty-three cardinals in all throughout the world—that His Holiness was close to death.

On October 6, Pius suffered a stroke, then, two days later, another one. Early in the morning of October 9, he died of a heart attack at the age of eighty-two. When the news of Pius's death came, Roncalli wrote in his journal, "Sister death came quickly and swiftly fulfilled her office. Three days [of illness] were enough. . . . One of my favorite phrases brings me comfort: 'We are not honored as museum keepers, but to cultivate a flourishing garden of life and to prepare a glorious future.' The pope is dead, long live the pope!"

On October 11, Roncalli offered a high requiem Mass for the departed pontiff and on October 12 headed off to Rome on the 9:40 train, bringing with him the *cappa magna,* the ceremonial red cape cardinals wore in those days for the sole purpose of welcoming a new supreme pontiff to his unique vale of tears.

PART II

The Soul of a Pope

CHAPTER SEVEN

Election and First Days
(October–December 1958)

Pope Pius XII had reigned for nineteen and a half years, though in the past several years he was slowed by failing health. It had been a momentous period in the history of the world and in the life of the Church. Among his many actions and achievements, Pius declared the dogma of the Assumption of the Blessed Virgin Mary, under the definition of papal infallibility put forth at the First Vatican Council in 1870 (the first and only time such a doctrine has been so defined) and canonized thirty-three new saints of the Church, including his predecessor, Pope Pius X.

"Pius' program represented a judicious synthesis of conservation and liberalism," according to Frank J. Coppa in *The Modern Papacy Since 1789*.

Although criticized as austere and authoritarian, Pius continued the papal commitment to social justice, writing in the first year of his pontificate [addressing his instruction to American bishops] that "the goods created by God for all men should in the same way reach all, justice guiding and charity helping."

In international affairs, the experienced pope-diplomat won some and lost some, clashing, for example, with the dictator of Argentina, President Juan Perón, on the issue of Church control of education in that country. The controversy over the pope's role in World War II—his portrayal as aloof from the reality of the mass extinction of the Jews of Europe—remains problematic, awaiting the opening of Vatican archives to resolve the historical and moral questions that have been raised about his policies toward the persecuted and the persecutors.

On December 8, 1945, Pius XII had promulgated a new constitution to govern the conclave, titled *Vacantis Apostolicae Sedis* (During Vacancies of the Apostolic See). Popes through the ages, as chief legislators of the Church, have often taken an active role in setting the rules and procedures that choose their successors. Even with periodic adjustments and innovations, the rules of the conclave then and now are largely the same as they have been for nearly ten centuries.

The document retained most of the regulations of Saint Pius X's similarly titled *Vacante Sede Apostolica* (On the Vacancy of the Apostolic See) of December 25, 1904, which dealt comprehensively with the election of the Roman pontiff and the role of the cardinals during the vacancy and in the electoral process. In title II, chapter I of the document, "The electors of the Roman Pontiff," the pope outlined the role of cardinals in selecting a new pontiff. Only they could cast votes. By the provisions of this constitution, even if the pope were to die while a council is being held in Rome or in any other place, the election was to be conducted by the cardinals alone and under no circumstance by the council, which would be suspended until the new pope decided to reopen it or not. A cardinal who had been excommunicated, suspended, interdicted, or subjected to any other ecclesiastical sanction was not excluded from the election. The censures were thus suspended, but only for the election.

Once a cardinal had been "created and published," he had the right to participate in the election even though he had not yet been invested with the traditional red hat of his rank. Cardinals who had been canonically deposed and those who, with papal consent, had renounced their dignity were not allowed to take part in the conclave. If a cardinal had not at least been ordained a deacon, he could not participate in the election unless he had a special privilege granted by the Roman pontiff.

After the death of the pope, cardinals were to await their absent colleagues, after which time they were ordered to enter the conclave and proceed to the election. Adjustments made by Pius XI in the motu proprio (a personal papal statement), *Cum proxime* (Concerning New Rules for the Election of the Pope) of March 1, 1922, ordered that the conclave should begin fifteen days after the death of the Roman pontiff. He also empowered the College of Cardinals to extend this period to eighteen days if they considered it necessary.

If a cardinal arrived after the beginning of the conclave but before a pope had been elected, he must be admitted into the conclave immediately. Unless legitimately impeded, as were Cardinals József Mindszenty and Alojzije Stepinac in 1958, all cardinals were obliged to participate in the election. If a cardinal refused to enter the conclave or left it after having entered, he would lose his vote and not be readmitted unless he was forced to leave because of sickness. All cardinals, if not impeded by sickness, were required to assemble for the ballot when the bell had sounded three times, and if a cardinal refused, he faced excommunication.

The most notable innovation introduced by Pius's constitution was the rule establishing that a cardinal could not be validly elected pope unless he obtained one vote over the traditional two-thirds majority. This new amendment precluded the possibility that this minimum could be obtained by the vote of the cardinal who received the necessary number of votes. This was the first time since Alexander III's 1179 constitution, *Licet de vitanda discordia* (which was a canon of the Third Lateran Council), that to be elected pope it was necessary to obtain more than two-thirds of the votes.

Pius XII had also created an unprecedented number of cardinals at two consistories: thirty-two on February 18, 1946, and twenty-four on January 12, 1953, including Roncalli, and a powerhouse "class" of new princes. The maximum for membership in the Sacred College of Cardinals had been set at seventy by Sixtus V in 1587 and confirmed in the Code of Canon Law of 1917. John XXIII would later increase the total number of cardinals, as would his successor, Paul VI. Today the canonical maximum is 120 eligible cardinal electors (with many more who survive past the age of eighty, when they become ineligible to vote in a conclave).

Reflecting the growing internationalism of the college, only seventeen of the fifty-one electors at the conclave were Italian, the lowest percentage (one-third) within a conclave since 1455, which resulted in the election of Callistus III as a compromise candidate. The next highest national representation was from France, with six cardinal electors. The United States boasted only two cardinals, Cardinal Francis Spellman of New York and Cardinal James Francis McIntyre of Los Angeles (a Spellman protégé), since Cardinal Edward Mooney of Detroit had died in Rome on the day of the conclave opening. Cardinal Celso Costantini, an Italian curial apparatchik, had died two weeks previously, and Cardinals József Mindszenty of Hungary and Alojzije Stepinac of Yugoslavia were not able to participate because they were stuck behind the Iron Curtain.

This was also the smallest number of electors since the election of Pope Pius VII in March 1800, by thirty-five cardinals gathered in a fourteen-week conclave in Venice under the protection of the emperor of Austria. At the time, Napoleon occupied Rome. In fact, the late Pius VI died in French custody in Valence, precipitating a political crisis within the Church.

Once before in the twentieth century, in 1903, a patriarch of Venice had been elected pope: Giuseppe Melchiorre Sarto, age sixty-eight. In 1978, another patriarch of Venice would be elected as Roman pontiff: the sixty-five-year-old Albino Luciani, who would reign for only thirty-three days in what came to be known as "the year of three popes."

After the 1903 conclave, in a prescription that presaged the historic papal election of 1958, the French Cardinal François-Desiré Matthieu stated, "We wanted a pope who had never engaged in politics, whose name would signify peace and concord, who had grown old in the care of souls, who would concern himself with government of the Church in detail, who would be above all a father and shepherd."

Roncalli was far removed from the *papabili,* the favored candidates for the papacy, but he had few, if any, real enemies. The slate of candidates was so unsettled that one Roman newspaper prepared biographies of more than twenty potential popes but did not include one for Roncalli.

On Saturday, October 25, 1958, at 6:08 P.M., the bell in the Court of Saint Damasus tolled three times, announcing that it was time to close the doors and seal the windows of the conclave for the election of the successor of Saint Peter. A few months earlier, the heads of the three orders of cardinals—cardinal bishops, Eugene Tisserant, dean of the College of Cardinals; cardinal priests, Josef-Ernest van Roey, archpriest; and cardinal deacons, Nicola Canali, archdeacon—along with the camerlengo, Aloisi Masella, had witnessed the barring of the two inside entrances of the area in the Court of Saint Damasus and the Borgia Court. Enrico Dante, as master of ceremonies, announced, *"Extra omnes"* ["Exit, all"], marking the official beginning of the conclave.

After the ceremonies ended, the cardinals retired to their cells. Cardinal Roncalli drew the cell set up for him in the offices of the Noble Guard. The sign, "Il Commandente," was still visible above the door. His seat in the Sistine Chapel, where the scrutinies would be held, was on the right of the main door, between Cardinal Valerio Valeri and Cardinal Gaetano Cigognani.

The next day, after Mass in the Pauline Chapel, the cardinals cast two votes in the morning and two in the afternoon. On the first scrutiny, or ballot, the relative positions of the candidates and the factions were revealed:

> Angelo Giuseppe Roncalli   20
> Gregory Peter XV Agagianian   18
> Valerio Valeri   4
> Giacomo Lercaro   4

Ernesto Ruffini   3

Giovanni Battista Montini   2

So, Roncalli won a plurality on the first scrutiny. It seemed that the cardinals might consider his age a benefit because he would probably not live long enough as pope to do any harm to the Church.

The second ballot was identical to the first. Outside, the world awaited the signal of the white smoke from the chimney above the Sistine Chapel, signaling the election of a new pope. After the first scrutiny, the smoke initially appeared white, before Vatican attendants added wet straw to the fire to make the smoke appear blacker on the horizon.

Two more full days of voting continued, with no one achieving the two-thirds needed for election. Roncalli lost support at one point, dropping as low as fifteen votes, behind Agagianian. The aged cardinal electors were reportedly tired and cranky. Then, key supporters lined up behind Roncalli as the balloting continued, including the curial lion, Alfredo Ottaviani, along with the Frenchman Tisserant, who had known the Bergamese cleric well and come to respect him tremendously in the postwar years.

The eleventh scrutiny was conclusive:

Angelo Giuseppe Roncalli   38

Gregory Peter XV Agagianian   10

Giacomo Lercaro   2

Valerio Valeri   1

At 4:50 P.M. on October 28, 1958, Angelo Roncalli was elected pope. For the first time, the patriarch of Venice had contravened the tradition that no major *papabile* ever recovers from a loss of support during the balloting. He had, instead, picked up just enough votes to achieve the throne.

The cardinals lowered the canopies above their seats in the Sistine Chapel: all but one, that of the newly elected pontiff. Cardinal James McGuigan of Toronto reached to his left and performed the courtesy for the late Cardinal Mooney of Detroit, whose body would be carried back from Rome to the United States within a few days.

Tisserant, the dean of the Sacred College of Cardinals, performed his ritual duty. He approached Roncalli to ask him the question required by canon law and the constitution governing conclaves: "Do you accept the election, canonically made, of yourself as pontiff?" In response, Roncalli drew from his pocket the Latin text over which he had worked through the long hours of the previous night and during the lunch hour in his cell:

At the sound of your voice, "I am made to tremble, and I fear." For what I know well of my poverty and insignificance is enough to bring me to confusion.

But seeing the votes of my brothers, the most eminent cardinals of the Holy Roman Church, the sign of the will of God, I accept the election made by them. I bow my head and my back to the chalice of bitterness and to the yoke of the cross.

On the solemn feast of Christ the King, all of us have sung: "The Lord is our judge; the Lord is our lawgiver; the Lord is our king. He will save us."

Usually the simple word, *Accepto,* is uttered at this time. Thus the new pope signaled that something different was going on here, as he did in response to Cardinal Tisserant's ritual ques-

tion, *"Quomondo vis vocari?"* (How do you wish to be called?).
His lengthy explanation of his choice of "John" caused even more
whispering comments from among the fifty cardinals who had
just elected him.

When elected pope, a man gives up his family and baptismal
name when he chooses a pontifical name. Angelo Roncalli had
explained to the cardinal electors immediately upon accepting his
election why he chose the name John:

> *Vocabor Johannes* (I wish to be called John). This name is
> sweet to us because it is the name of our father. It is sweet
> to us because it is the name of the humble parish church
> in which we were baptized; it is the name of innumerable
> cathedrals scattered throughout the world and first of all the
> sacred Lateran Basilica, our cathedral.
>
> It is the name which has been borne by more popes in
> the long list of Roman pontiffs. In fact there are 22 supreme
> pontiffs with the name of John of undoubted legitimacy.
> Practically all have had a short pontificate. We have pre-
> ferred to cover the littleness of our name behind this mag-
> nificent succession of Roman pontiffs.
>
> But we love the name of John, so dear to us and to the whole
> Church, especially because of the two who have born it, the two
> men, that is, who were closest to Christ the Lord, the Divine
> Redeemer of the whole world and founder of the Church.
>
> John the Baptist, the forerunner of our Lord, was not the
> light himself, but a witness to the light, an invincible wit-
> ness to truth, justice, and freedom, in his preaching, in his
> baptism of penitence, and in the blood which he shed.
>
> And the other John, the disciple and evangelist, beloved
> by Christ and his dearest mother, who at the Last Supper

leaned on the breast of the Lord and drew from thence that charity of which he was a living and apostolic flame until the end of his ripe old age.

May John the Evangelist who, as he himself relates, took to himself Mary the Mother of Christ and our mother, support together with her this exhortation, which is meant for the life and joy of the Catholic and apostolic Church, and also the peace and prosperity of all nations.

My little children, love one another; love one another because this is the great commandment of the Lord.

Immediately afterward, the master of ceremonies drew up the act of acceptance, with the secretary of the conclave signing as witness. Then John XXIII, as he was now styled, proceeded to the sacristy of the Sistine Chapel to don the papal vestments. As he did so, he took his red zucchetto, or skullcap, and placed it on the head of the secretary of the conclave, Albert di Jorio, as a sign that he would be made a cardinal in the first consistory of the new reign. This revived an age-old custom that had lapsed since Pius X had done the same for Raffaele Merry del Val back in 1903.

The robing of the new pope involved many new vestments, namely, the white cassock with stockings of the same color, the red shoes with a cross of gold on top, the rochet (the knee-length lace vestment similar to a surplice), red mozzetta, red stole, and white zucchetto.

John's conclavist, his designated assistant during the conclave, Monsignor Loris Capovilla, his priest-secretary in Venice, was called into the sacristy, still ignorant of the election. It was traditional for the conclavist of the newly elected pope to assist him in vesting.

Famously, the papal tailor, Annibale Giammarelli, provided

three complete sets of vestments to ensure that one would fit the new pope, no matter who might be elected. The first set John tried on could not even be buttoned over his 200-pound girth. The second was a little better, and it was in these robes that he spent the balance of the first day of his pontificate. The following day, Capovilla telephoned the clothier to complain about the bad fit. Giammarelli asked whether John had tried all three outfits and was told there had been time for only two robings. "Try the third," he said. When they did, they found the third set fit John perfectly, for Giammarelli had anticipated even before the conclave that John could very well be the next pope and had prepared one set of vestments for him.

After accepting the obeisance of his brother cardinals, John proceeded to the balcony above the piazza to deliver his first blessing as Roman pontiff *urbi et orbi* (on "the city and the world"). The moment was captured on television for the first time in history, and it is estimated that an astonishing billion people saw or heard the announcement of the newest supreme pontiff's election.

His pontificate of 1,680 days began with a bang: on November 2, a mere five days after his election and just two days before he would be crowned with the triple tiara, Pope John XXIII floated the idea of convoking the first ecumenical council of the Church in ninety years with Cardinal Ernesto Ruffini. Further, he decided to create twenty-three new cardinals in a consistory to be held on December 15, breaking the limit of seventy established in 1587 by Pope Sixtus V.

The College of Cardinals, which would swell to a membership of some 200 in the second decade of the twenty-first century (though a new maximum of only 120 under the age of eighty would be eligible to vote in conclave), would never be the same again. Nor would the papacy itself, thanks to this new old pope.

CHAPTER EIGHT

The First Year
(January–December 1959)

Pope John XXIII wasted no time. Though not known as a frenetic doer, the newly elected and crowned pontiff entered the first year of his reign with a full agenda.

The story, as John later told it, was that he and his secretary of state, Cardinal Domenico Tardini, met on the afternoon of January 20, 1959, three months after John had been elected pope, to discuss certain matters concerning the direction in which John wished his papacy to proceed, particularly in light of the fact that the world "was plunged into so many grave anxieties and troubles." Suddenly, as John later expressed to group of Venetian pilgrims to the Vatican in May 1962, "an inspiration sprang up within us as a flower that blooms in an unexpected springtime. Our soul was illuminated by a great idea. . . . A word, solemn and binding, rose to our lips. Our voice expressed it for the first time—a council."

This anecdote shows that even popes are human and can be prone, like the rest of us, to embellishment. John's diary, written on the evening of the day in question, is more matter-of-fact:

In conversation with Tardini, Secretary of State, I wanted to test his reaction to my idea of proposing the project of an Ecumenical Council to the [cardinals] of the Sacred College when they met at Saint Paul's on the 25th of this month. . . . I was rather hesitant and uncertain. Tardini's immediate response was the most gratifying surprise that I could have expected: "Oh, that really is an idea, an enlightening and holy idea. It comes straight from heaven, Holy Father."

Tardini himself noted, in a shorthand memo written that night, that the pope had mentioned three initiatives to him: "Roman Synod. Ecumenical Council. *Aggiornamento* [updating] of the Code of Canon Law."

There is obviously a great difference between being struck by divine lightning and "testing" Tardini's reaction—as there also is between "splendid initiatives" and an idea that comes "straight from heaven." But this disparity in stories—with the initial recountings no doubt being the correct ones—only underscores how controversial an idea the ecumenical council really was. Not only was John cautious in bringing the council up in front of the crusty and conservative Tardini, but, even after three years of preparation, as Vatican II was only months away, he was already mythologizing its inception for the world at large.

And John was right to be cautious. On Sunday, January 25, five days after he told Tardini of his decision, he spoke to a group of seventeen influential cardinals after attending Mass at the Basilica of Saint Paul's-Outside-the-Walls during the final day of the celebration of Christian Unity. Tardini had told some of the cardinals what was coming; others were unaware. John began by telling the group that he had come to a momentous decision. There were periods in Church history, he said, when the Church sought "greater clarity of thinking," as well as "a strengthening of the bond of unity, and greater spiritual fervor." With these three goals in mind, he then said to the cardinals, "Trembling with emotion and yet with humble resolution, we put before you the proposal of a double celebration: diocesan synod for Rome and an ecumenical council for the universal Church."

John then looked at the assembled and said, "I would like to have your advice." The cardinals simply stared at him, without a word. John was sorely disappointed. As he wrote later, "Humanly, we could have expected that the cardinals, after hearing our pronouncement, would have crowded around to express approval and good wishes." Instead, his grand plans met with a lengthy, stony silence. There were any number of reasons for this. Some of the cardinals were shocked that a pope whom many saw as merely a "transitional" figure, holding down the fort, as it were, until a more dynamic Church leader might come along, was proposing something so monumental. After all, there were only twenty such councils in Church history; the last had been the First Vatican Council (so-called because it took place within the Vatican) in 1870.

But the main reason that the cardinals withheld their approval was that they were members of the Roman Curia. Withholding approval was, in a sense, what they did best. The Curia—*curia* from the Latin word for "court"—is the administrative arm of the Holy See. They quite literally run the departments, or congregations, of the Church on behalf of the pope. Pontiffs came and pontiffs went, but the Curia lived on. The cardinals had a vested interest in protecting the status quo.

John was acting in an entirely unexpected way—at least to them.

From the very beginning of his papacy, John had given signals that he saw himself as a pastoral pope, a benevolent father figure to the Universal Church. Many of his admirers used the term *collegial* to describe him. Without giving up his traditional authority as pope, he sought consensus. And his was always an ecumenical reach—as his pre-papal relations with the Eastern Orthodox and Anglican churches, and with the Jewish faith, had proven.

One other thing unsettled the cardinals. In the past, the great councils had been called to condemn apostasy. The Council of Nicea in 325, for instance, was called to put an end to Arian heresy. The Council of Constance, held between 1414 and 1418, dealt with the Avignon popes and the papal schism. The Council of Trent, which continued on and off for eighteen years during the late 1500s, condemned the Protestant Reformation. And the yearlong Vatican I, called by Pope Pius IX in 1869, declared the infallibility of the pope in matters of faith and morals—a pronouncement designed to reassert the Church's authority in a world awash in rationalism and Modernism. (Vatican I was cut short when the Franco-Prussian War broke out and the kingdom of Italy captured Rome, forcing Pius to suspend the council. It was never reopened.)

The Curia knew the history of the Church councils well. And they saw that what John was proposing was a pastoral council, a council in which no heresy would be expunged, no dogma reasserted. While the direction the council might take was unclear, on that cold January day, it was clear that anything could happen. And the Curia, as well as other cardinals of the Church, did not like that at all. Cardinal Francis Spellman of New York said, in public, "I do not believe that the pope wanted to convoke a council, but he was pushed into it by people who . . . misconstrued what he said."

Spellman's critique fell far from the mark—and it reflected the error that so many of John's colleagues and superiors had made over many years: underestimating the peasant-born cleric. Also, it was not immediately apparent to most that a turning point in the history of the Church had been reached. A new dynamic, inspired by the Holy Spirit (according to the faithful), was at work in the old Church, a spirit of reform and renewal.

Archbishop Giacomo Lercaro of Bologna went further: "How could he have dared to convoke a new council after one hundred years and within less than three months after his election? . . . Either Pope John has been rash and impulsive, with a lack of breeding and experience . . . or else in actuality Pope John has done this with calculated audacity, though obviously not capable of foreseeing all the details. . . ."

Even Giovanni Montini, friend and counselor of John, whom the pope had made a cardinal less than three weeks after his election, shook his head as he memorably told a friend, "This holy old boy doesn't seem to realize what a hornet's nest he's stirring up." The council, Montini warned, would unleash "expectations, dreams, curiosity, utopias, velleities of every kind and countless fantasies."

Yet Montini of Milan, who had received the red hat and the title of SS. Silvestro e Martino ai Monti on December 18, 1958, as the first new cardinal created by John—and who would inherit the Second Vatican Council as Pope Paul VI—knew the pope's heart. As the council was announced, "a flame of enthusiasm swept over the whole Church," the cardinal later wrote. "[Pope John] understood immediately, perhaps by inspiration, that by calling a council he would release unparalleled vital forces in the Church."

Perhaps John wasn't really embellishing at all when he wrote of his conversation with Tardini that day. His *was* an inspiration, born of the Holy Spirit, that illuminated not only his whole soul, but the Church as well.

Before the council started, however, there was a good deal of work to do.

Issued on June 29, 1959, the Feast of the Apostles Peter and Paul, Pope John's first encyclical, *Ad Petri cathedram* (To the Chair of Peter) was devoted to truth, unity, and peace, the three-legged program of his papacy. Seen in the context of its time, the encyclical is part of John's history of meeting the secular world head on.

At its heart, it is three things: a promotional piece for the forthcoming ecumenical council, a protective encouragement for the Church behind the Iron Curtain, and an extended hand, if not a proffered embrace, to the estranged churches of Christendom.

John previewed his encyclical at a vespers service at Saint Peter's Basilica on the eve of the Memorial of Saints Peter and Paul. The occasion made sense. Peter had been martyred under Rome's Nero and Paul was the Church's proto-missionary supreme, and the pope was concerned with Cardinal József Mindszenty of Hungary and Cardinal Alojzije Stepinac of Yugoslavia, who were up against communist pressure. He spoke to a crowd of several thousand from the place below which tradition holds Peter is buried and lamented that officials in some communist countries, such as in Hungary, were fostering discord in the Church.

The vespers service drew eighteen cardinals, underscoring the importance of the pope's preview of his encyclical. The text of his address was made public through the United Nations, which designated the year in an attempt to alleviate the suffering of

15 million refugees worldwide. This in itself can be deemed significant. It came a day after a similar appeal by the World Council of Churches. In these actions and with this timing, one can see the kernel of the universal and urgent embrace we find later in *Pacem in terris*. In the pope's message on refugees, he spoke words that would resonate throughout his pontificate:

> What kind-hearted man could remain indifferent to that sight: so many men, women and even children deprived, without any fault of their own, of some of the fundamental rights of the human person: families divided up in spite of their own wishes, husbands separated from their wives, children kept away from their parents.
>
> What a sorrowful anomaly in modern society, so proud of its technical and social progress! Everybody has the duty to take this matter to heart and to do whatever is in his power in order to bring this sad situation to an end.

In some circles, this aspect of the pope's public comments drew even more attention than the letter itself. In a story headlined "Pope Says Reds Martyr Church," the *New York Times* centered its coverage on the pope's remarks during a sermon on the Feast of Saints Peter and Paul, not at all on the encyclical. Speaking to 10,000 people in Saint Peter's and separately to 20,000 in the square, Pope John had brought up the Church of Silence, behind the Iron Curtain, twice in a little more than twelve hours, according to one account.

But what did *Ad Petri cathedram* itself have to say on this subject? It devotes several impassioned paragraphs to the topic but certainly does not lead with this issue. "Many of these sublime words apply in a special way to those who are members of the

'Church of Silence,' for whom we are all especially bound to pray to God," the pontiff declared, recalling even more pronouncements he had made on the subject weeks earlier, on Pentecost Sunday and on the feast of the Sacred Heart of Jesus. But a more detailed, and more nuanced, discussion is found earlier in the encyclical, under the subheading "The Church Persecuted" in the Vatican's rendering:

> 137. We have exhorted all our children in Christ to avoid the deadly errors which threaten to destroy religion and even human society itself. In writing these words our thoughts have turned to the bishops, priests, and laymen who have been driven into exile or held under restraint or in prison because they have refused to abandon the work entrusted to them as bishops and priests and to forsake their Catholic faith.
>
> 138. We do not want to offend anyone. On the contrary, We are ready to forgive all freely and to beg this forgiveness of God.
>
> 139. But We are conscious of our sacred duty to do all that We can to defend the rights of our sons and brethren. Time and time again, therefore, we have asked that all be granted the lawful freedom to which all, including God's Church, are entitled.

Typical of the style and substance of John, even early in his pontificate, here he is always eager to extend an olive branch, even as he steadfastly affirms rights and principles of the Church.

What of the rest of *Ad Petri cathedram*? To our ears, its language may seem stiff and formal. But even in this inaugural circular letter we hear some of the paternal voice that would become more characteristic of the Good Pope. He uses the royal "we," but he cannot seem to hide his gentle, personal demeanor. Indeed, *TIME* magazine noted the document's "kindly" and "fatherly" approach.

Yet some critics saw John as a writer still trying to find his voice, as someone too overshadowed by the voices of Tardini and others. Indeed, some observers found the use of terms like *return,* calling other Christians back to Rome, offensive and condescending.

Quite explicitly and didactically, the pope says his writing is about truth, unity, and peace. Then, as a lecturing professor might, he proceeds to explicate on each of these focal points. As for truth, he defends the Roman Catholic Church, decrying the view that "one religion is just as good as another." He also devotes significant attention to the rights and responsibilities of mass media, quoting Leo XIII, which he does often in this encyclical, to condemn works that "mock virtue and exalt depravity." As an unintended but striking coincidence, on this same day, June 29, the U.S. Supreme Court overturned a New York State ban on the movie version of D. H. Lawrence's *Lady Chatterley's Lover*—a landmark decision that arguably changed American culture. And in what was an almost expected reaction, the pope praises advances in modern science but cautions to what end and purpose. "But why do we not devote as much energy, ingenuity, and enthusiasm to the sure and safe attainment of that learning which concerns not this earthly, mortal life but the life which lies ahead of us in heaven?"

In the tradition of popes before him, in particular Leo XIII, the pontiff links social justice to unity and peace. He goes so far as to reassert Leo's and Pius XII's view of classes as a structure commanded by God to rebuff Marxist tenets. But when it comes to peace, John's words ring with a new sense of ardor, passion, and mercy. "God created men as brothers, not foes," he says. "We are called brothers. We actually are brothers." Foreshadowing Paul VI's famous declaration, "No more war; war never again," before the United Nations, he pleads, "There has already been enough warfare among men! Too many youths in the flower of life have shed their blood already! Legions of the dead, all fallen in battle, dwell within this earth of ours. Their stern voices urge us all to return at once to harmony, unity, and a just peace."

In many ways, the heart of John's message is unity: unity of and within the Catholic Church, unity with separated Christian brethren, and the unity of the human race. He aligns these goals with the ecumenical council that will follow. Regarding the Church, he speaks of three types of unity within it: unity of doctrine, unity of organization, and unity of worship.

In making a passionate invitation to Christian unity, John offers a powerful quote from Saint Augustine: "Whether they wish it or not, they are our brethren. They cease to be our brethren only when they stop saying 'Our Father.'" And even more tenderly and ardently, he says to all who are separated from the Chair of Peter, "I am . . . Joseph, your brother," recalling the touching Genesis story of exile and reunification.

Remarkably, he cites Cardinal John Henry Newman to make a case for doubt and uncertainty—something his successors in the twenty-first century would be unlikely to do. He even quotes what he calls a common saying from Saint Augustine: "In essentials, unity; in doubtful matters, liberty; in all things, charity." It's pure John, and not seen since.

In twenty or so paragraphs, John offers encouragement, guidance, and love specifically to ranks within the Church: bishops, clergy, religious men and women, missionaries ("ambassadors of Christ," about whom he promises to say more—which he did later with *Princeps pastorum*—Prince of the Shepherds, or of the Pastors), and laypeople, especially Catholic Action.

Still, one cannot ignore the prodding and advocating and fostering of the Vatican Council espoused in his first encyclical. This man of deep faith showed that characteristic when he said, "The outcome of the approaching Ecumenical Council will depend more on a crusade of fervent prayer than on human effort and diligent application."

Issued only two months after John's first encyclical, *Sacerdotii nostri primordia* (The Beginning of Our Priesthood) uses the one hundredth anniversary of the death of Saint John Marie Baptist Vianney, the Curé of Ars, to guide, inspire, and even challenge Roman Catholic priests.

Saint John Vianney (1786–1859) was a country priest from the Lyon area, in the Rhone Valley of France. John had a special affection for the Curé of Ars because, as noted in *Sacerdotii nostri primordia,* the saint's life intersected at many points with John's own spiritual journey. Vianney was declared Blessed on January 8, 1905, just months after John's ordination on August 10, 1904. Pope John and Saint John Vianney shared other milestones as well. John learned that his own patron and mentor, Giacomo Radini-Tedeschi, was appointed a bishop on the same day John Vianney was named Blessed, that is, beatified. Shortly thereafter, Father Roncalli became Radini-Tedeschi's assistant, and early in 1905, the young Roncalli made a pilgrimage to the tiny village of Ars, in France, to pay homage to the subject of his future encyclical. Then, in March 1925, Father Roncalli became a bishop, just a few months before the Curé of Ars was named a saint and the patron of priests of the Roman Catholic Church.

In terms of its substance, the encyclical easily could have been written by Pope Benedict XV, or any pope before John. As a matter of pure speculation, what would this encyclical have looked like if John had outlived Vatican Council II? Would it not have a different tone and substance? What might have been the impact on this encyclical of a key by-product of the council, the Dogmatic Constitution on the Church, or *De Ecclesia*? John's encyclical focuses on priestly virtue, making no mention of "the royal priesthood," the Church's baptized lay members on which Vatican II cast a spotlight, rather than focusing exclusively on the Church's clergy and hierarchical structure. This paragraph from the encyclical would seem out of place, almost insulting, if it were written in, say, 1965:

> If there were no priests or if they were not doing their daily work, what use would all these apostolic undertakings be, even those which seem best suited to the present age? *Of what use would be the laymen who work so zealously and generously to help in the activities of the apostolate?* [Emphasis added.]

To make a broader and more extreme leap, imagine, in light of his tender guidance of his priests, how disheartened the Good Pope would be to learn of the abuse scandal that rocked the Church in the 1990s and beyond. A naïve and simplistic conclusion would be: This would have never happened if priests took this letter as seriously as it was intended to be.

Picture John's words from this encyclical imparted to priests and bishops in the context of the abuse scandal:

> And so We do not hesitate to speak to all of these sacred ministers, whom We love so much and in whom the Church rests such great hopes—these priests—and urge them in the name of Jesus Christ from the depths of a father's heart to be

faithful in doing and giving all that the seriousness of their ecclesiastical dignity requires of them.

Dated August 1, 1959, the 10,000-word document has three main parts. After the introduction recalling his fond personal remembrances of Saint John Vianney, the first section of the encyclical letter discusses Vianney's legendary asceticism. Vianney was known for fasting and self-denial in the extreme. He ate and slept little. When he did sleep, it was often on the floor. He was, as Pope John recalled, "hard on himself, and gentle with others." In painting a portrait of Saint John Vianney, the pope discourses on the so-called evangelical virtues of poverty, chastity, and obedience, which typically govern those in religious orders. Despite noting that diocesan priests are not formally bound by these vows, John remarks that "these counsels offer them [churchmen] and all of the faithful the surest road to the desired goal of Christian perfection."

Much is made of Vianney's personal, Franciscan-style poverty and compassion for the poor. John frequently cites the curate's own words, for example: "My secret is easy to learn. It can be summed up in these few words: give everything away and keep nothing for yourself." One can already begin to see a consistent strain of thought in John's papal pronouncements: here we see him extolling the parish priest who freely accepted beggars—and later we see this expanded to a macro scale in encyclicals such as *Mater et magistra* (Mother and teacher) and *Pacem in terris* (Peace on earth).

As the pontiff holds up Vianney as a paragon of chastity, today's readers might find it curious that the word *celibacy* never enters the discussion. This was not in the least unusual in 1959, and yet how much the winds of change that John himself set in motion would alter the conversation in just a few short years, owing to the debates of the ecumenical council.

He also celebrates Vianney as an exemplar of priestly obedience, especially to his religious superiors. The pope recalls that Vianney had always wanted to lead a more monastic and solitary life—he pined for it—and yet Vianney obeyed his superiors and remained a curate, serving his country parish in France. His wish was never granted; he obeyed. Naturally, as the supreme pontiff, it made organizational sense for Pope John to reinforce this tenet. Yet if we read it today, and after decades of sometimes fractious debate within the Church, one can only surmise: Is this the hallmark of an era never to be seen again in the Church, despite the rigorous efforts of recent popes? Or is it a model that is increasingly embraced again as a path to priestly holiness? And what role did, or could have, obedience played—to good or ill effect—in the scandals that have plagued the Church?

The encyclical's second segment focuses on Saint John Vianney as a model of priestly holiness, chiefly through prayer—constant prayer. The pope underscores the necessity of a priest's prayer life, listing prayerful practices as diverse as recitation of the Rosary, praying the Holy Office, meditation, visits to the Blessed Sacrament, examinations of conscience, and most important, devotion to the Eucharist and the Holy Sacrifice of the Mass. Regarding the Mass, Vianney—and hence, Pope John—goes so far as to see external and internal attentiveness and piety in celebrating Mass as keys to a holy priesthood, warding off flat and stale vocations.

The third part of *Sacerdotii nostri primordia* gets up close and personal, if you will, taking a close look at the pastoral zeal of this good shepherd of souls. Undoubtedly drawing parallels to his own secular era—and ours, if we read the encyclical today—John noted that Vianney had his own parish challenges. In an era after the French Revolution, his flock was in sore need of conversion. Nowhere was Vianney's pastoral passion embodied and shown more than in the confessional. It is said he heard confessions daily for thirty years. The pope rattled off stupefying statistics, saying that Vianney was known to have heard confessions fifteen hours a day, from early morning into the night, to the tune of 80,000 confessions a year! This ministry of mercy surely was not merely numerical but rather a stunning example of mercy and anguish for sinners and limitless compassion for his flock. It was said that Vianney's penances were light; he'd bear the burden, essentially, for the poor sinners whose confessions he heard. "I impose only a small penance on those who confess their sins properly; the rest I perform in their place."

Other priestly virtues that John inventories are preacher, catechist, teacher, and learner. Regarding the role of catechism, John hearkens back to the Council of Trent, saying that the council "pronounced this [the priest's role of catechist] to be a parish priest's first and greatest duty." And as for preaching, John declares that "even Saint Francis de Sales would have been struck with admiration" for Vianney's preaching. Here again, John may have even unwittingly seen parallels with himself: not the world's most polished speaker, but one who won souls by his humility, honesty, and personal holiness.

In John's last major verse in his hymn to priestly excellence, he asks that the faithful pray for priests and that families support vocations to the priesthood. "We have complete confidence that the young people of our time will be as quick as those of times past to give a generous answer to the invitation of the Divine Master to provide for this vital need. . . . So let Christian families consider it one of their most sublime privileges to give priests to the Church;

and so let them offer their sons to the sacred ministry with joy and gratitude."

We can only speculate how the pope might have recast his message today to address critical priest shortages.

What seems unmistakable in all this is the influence that Saint John Vianney had on Pope John's spiritual formation: his personal holiness, his priesthood—and his pontificate. We have every right to conclude that Pope John took Saint John Vianney's words and practices to heart and embraced them in his own ministry, a ministry that ultimately embraced the world.

In *Princeps pastorum,* published November 28, 1959, John XXIII envisioned the growth of the Church in formerly colonial lands that were awakening to independence, often in volatile fashion. He intuitively understood where population numbers would increase and where the Church's most fertile grounds would be: not in the Europe of old but instead in the developing nation-states of Africa, Latin America, and Asia. It is noteworthy that the encyclical's publication coincided with the creation of eight archdioceses with twenty-nine suffragan, or subordinate, dioceses in West Africa (Belgian Congo and Ruanda-Burundi). And of the new bishops named, five of them were native Africans.

Issued on November 28, 1959, thirteen months to the day after John's election, *Princeps pastorum* is Pope John's encyclical letter on the Church's missions. As with many papal encyclicals, his fourth encyclical commemorates an earlier letter, in this case *Maximum illud* (On the Propagation of the Catholic Faith Throughout the World), from Pope Benedict XV, in 1919, forty years earlier almost to the day. In *Maximum illud,* Benedict, who was writing just after World War I, asserted the importance of native clergy in mission lands and underscored the importance of evangelization.

Princeps pastorum gets its name, as is customary, from its open-ing words in Latin. In this case, John borrows from a biblical ex-hortation found in the first letter of Peter in the context of Church leaders being urged to tend to the flock of the Good Shepherd: "And when the chief Shepherd is revealed, you will receive the unfading crown of glory" (1 Peter 5:4).

Among highlights of the text are the following: a call for not only a native clergy but also local hierarchies, which he did on that day for West Africa; the need for personal sanctification, a com-ment likely intended to thwart local customs that flouted conven-tional Western morals; native teachers in seminaries; adaptation to local customs and cultures; promotion of social welfare and material improvement projects; condemnation of ultranation-alism; establishment of missiology as part of the curriculum of seminaries in mission lands; call for lay help and from the Church at large; and encouragement for public defense of Christian life and against persecution.

Perhaps drawing battle lines for the liberation theology debate decades later, he declared:

> Therefore, in mission territories, the Church takes the most
> generous measures to encourage social welfare projects,
> to support welfare work for the poor, and to assist Chris-
> tian communities and the peoples concerned. Care must
> be taken, however, not to clutter and obstruct the apostolic
> work of the missions with an excessive quantity of secular
> projects. Economic assistance must be limited to necessary
> undertakings which can be easily maintained and utilized,
> and to projects whose organization and administration
> can be easily transferred to the lay men and women of the

particular nation, thus allowing the missionaries to devote themselves to their task of propagating the faith, and to other pursuits aimed directly at personal sanctification and eternal salvation.

At the outset of *Princeps pastorum,* John refers to his own early experience.

Our zeal for the Pontifical Congregation for the Propagation of the Faith, a function which we most willingly performed during four years of our priestly life. We happily recall Whitsunday in 1922, the third centenary of the foundation of the Congregation for the Propagation of the Faith, which is especially entrusted with the task of carrying the beneficial light of the Gospel, and heavenly grace, to the farthest reaches of the earth. It was with great joy that we participated in the Congregation's centennial festivities on that day.

His experience in this role shaped his thinking and provided an emotional attachment and zeal for this element of his pontificate. Furthermore, his postings in Bulgaria, Turkey, Greece, and France added to his deep recognition of the Church's universal outreach and dedication to missiology.

During the Cold War, the Church often confronted opposition, or outright persecution, in communist nations. In fact, on the day of the encyclical's publication, Fidel Castro in Cuba openly criticized the National Catholic Congress meeting in Havana as an attack on the Cuban revolution. The Church also faced harsh opposition in eastern bloc Soviet satellite countries and in China. The encyclical's several pointed references to the

Church's challenges likely were not-too-veiled references to these conditions.

> There appear before our eyes other regions of the world
> where bountiful crops grow, thrive, and ripen, or regions
> where the labors of the toilers in God's vineyard are very
> arduous, or regions where the enemies of God and Jesus
> Christ are harassing and threatening to destroy Christian
> communities by violence and persecutions, and are striv-
> ing to smother and crush the seed of God's word. We are
> everywhere confronted by appeals to us to ensure the eternal
> salvation of souls in the best way we can, and a cry seems to
> reach our ears: "Help us!" Innumerable regions have already
> been made fruitful by the sweat and blood of messengers of
> the Gospel.

In the encyclical, John notes that on October 10, 1959, he had received more than 400 missionaries at Saint Peter's Basilica at the Vatican, giving each one a crucifix "before leaving for distant parts of the world to illumine them with the light of Christianity." He termed it one of the happiest events of his pontificate thus far.

We can view John's exhortation as a significant part of a long continuum of the Church's mission work, part of a long tradition of teaching on missiology. Of course, reaching back, we have great examples in Peter as shepherd and Paul of Tarsus as the foremost evangelist of the Gospels. Over the centuries Francis Xavier, Matthew Ricci, and others stood out as missionary exemplars. (The pope even singled out Ricci as a model missioner, educating native citizens of mission lands.)

As for papal documents, *Princeps pastorum* is part of a series of pronouncements. These include, as noted earlier, *Maximum illud* in 1919, but also *Rerum Ecclesia* (On Catholic Missions) by Pius XI in 1926; *Evangelii praecones* (On Promotion of Catholic Missions) by Pius XII in June 1951; *Fidei donum* (The Gift of Faith), again by Pius XII, in April 1957; *Evangelii nuntiandi* (On Evangelization in the Modern World) by Pope Paul VI in December 1975; and *Redemptoris missio* (Mission of the Redeemer) by Pope John Paul II in December 1990. And many of these concepts came to full flowering in Vatican II documents such as *Lumen gentium* (Light of Nations) and *Ad gentes* (To the Nations), which showed respect for local customs and cultures in mission lands.

This impulse had a rich history in the Church. Pope Gregory the Great (590–604), writing to Abbot Mellitus, fellow missionary of Saint Augustine of Canterbury, asked Augustine to purify pagan temples with holy water and to place relics of saints in altars, but not to destroy the local temples. And in 1659, the Sacred Congregation of the Propagation of the Faith gave the following instruction to vicars apostolic of foreign missions:

> Do not in any way attempt, and do not on any pretext persuade these people to change their rites, habits and customs, unless they are openly opposed to religion and good morals. For what could be more absurd than to bring France, Spain, Italy or any other European country to China?

Pope Pius XII's *Evangelii praecones* forcefully advocated retaining local culture as part of missionary work:

> Another end remains to be achieved, and we desire that all should fully understand it. The Church from the beginning down to our time has always followed this wise practice:

let not the Gospel, in being introduced into any new land, destroy or extinguish whatever its people possess that is naturally good, just or beautiful. For the church, when she calls people to a higher culture and a better way of life under the inspiration of the Christian religion, does not act like one who recklessly cuts down and uproots a thriving forest. No, she grafts a good scion upon the wild stock that it may bear a crop of more delicious fruit.

Although some consider Pope Benedict XV's *Maximum illud* as the "Magna Carta of modern Catholic missiology," John built upon this and modernized it, as it were, and put it into action in Africa, for example. John also broke from the past in his viewpoint and language. Gone, for the most part, were the Eurocentric perspectives and the use of condescending terms such as "savages," "uncivilized," "barbarous peoples," "heathens," "infidels," "pagans," and "pagan nations" found in earlier papal documents.

In fact, one can say that we were just beginning to see glimpses of the universal embrace that would find fruition in *Pacem in terris*. In this encyclical, we find this perspective—explicit in ways that previous papal language tended to avoid—in this passage in which the pope quotes an address of his own in the summer of 1959 to participants of the Second World Congress of Negro Writers and Artists:

We ourselves have already expressed our thoughts on this matter as follows: "Wherever artistic and philosophical values exist which are capable of enriching the culture of the human race, the Church fosters and supports these labors of the spirit. On the other hand, the Church, as you know, does not identify itself with any one culture, not even with

European and Western civilization, although the history of the Church is closely intertwined with it; for the mission entrusted to the Church pertains chiefly to other matters, that is, to matters which are concerned with religion and the eternal salvation of men. The Church, however, which is so full of youthful vigor and is constantly renewed by the breath of the Holy Spirit, is willing, at all times, to recognize, welcome, and even assimilate anything that redounds to the honor of the human mind and heart, whether or not it originates in parts of the world washed by the Mediterranean Sea, which, from the beginning of time, had been destined by God's Providence to be the cradle of the Church."

Princeps pastorum certainly did not get the sort of media attention afforded the later landmark encyclicals *Mater et magistra* and *Pacem in terris.* Although the *New York Times* gave front-page coverage, it did not provide a complete text, as it did with *Pacem.* The U.S. Jesuit periodical *America* devoted only three paragraphs in its December 12, 1959, issue, saying, "It is a fact of great significance for the future of the Church that the areas of its greatest missionary concern happen to be also the most crucial areas in contemporary world politics." *America* went on to assert that "the large space given to the role of the laity is perhaps the distinctive aspect of *Princeps pastorum.* . . . The new encyclical seems to have conferred new stature and importance on the lay apostolate not only in the missions but everywhere."

This encyclical garnered little or no comment from *National Review, Commonweal,* and American weekly newsmagazines like *TIME* and *Newsweek.*

Within the first year of his reign, in these three papal documents, Pope John articulated the themes of his papal ministry and of the council to come: engagement with the secular world, understanding the ordained priesthood as a pillar of the Catholic faith, and evangelizing—spreading the ancient faith—everywhere on earth. The still-new pope had staked out the landscape upon which he would begin to erect his dreamed-of edifice: the Vatican Council.

CHAPTER NINE

A Unique Pontificate
(January–December 1960)

The papacy then, a half-century ago, was different in significant ways than today's office, though the underlying theology and outward emblems of the institution have remained more or less constant for more than a millennium. John XXIII was, to a great degree, responsible for the reforms that took place during the decades since his pontificate: within just a few years Paul VI began to streamline the organization of the Vatican agencies and to expand the membership in the College of Cardinals; and John Paul II and Benedict XVI continued the process of reform. For example, the latter eliminated the title Patriarch of the West from the list of pontifical epithets that the pope had carried on his head (like the triple tiara that Paul put aside) for many hundreds of years.

In John's time, the 108.7-acre Vatican City State, the smallest country in the world with a population less than the total membership in the U.S. Congress, was run by men of Italian origin, exclusively, and its official language was Latin. It remains a closed system of government today, though some moves toward more relative transparency have been instituted in the twenty-first century.

The head of the Vatican City State, the pope, is the country's chief executive, the chief legislator, and the chief judge all in one. The nation has its own postage stamps and issues its own coins, yet it uses the euro (formerly the Italian lira) as legal tender and depends upon the government of Italy to transport its airmail. There are no street addresses in Vatican City, but postal delivery personnel know where everyone lives and works. In John's day, Vatican coins, which were the same size as Italian coins, had the pontiff's head engraved on them and usually bore a motto such as "This is the root of all evil" or "It is better to give than to receive."

The Vatican flag, then and now, consists of two equal vertical stripes of gold and white with the papal tiara above two crossed keys on the white stripe. John had ten or so private cars that were parked in the Apostolic Stable, which was once used for papal horses. In his time there were six gasoline pumps in the Vatican, all of them carrying the same brand name, prominent in its time: Esso.

Many of the citizens of Vatican City, none of whom was subject to Italian income taxes (only an annual Vatican tax of 300 lire, about 50 cents), lived in Italy rather than on Vatican ground. Vatican gates closed at 11:30 P.M. A resident who wished to go out to dinner or the opera was required to get special permission and make special arrangements to return to the city (country) after the closing of the gates. An "alien" (non-Vatican citizen) who was a guest for dinner at a Vatican apartment had to leave the tiny nation before the frontier shut down.

Although most prices within the Vatican were comparable with those of the neighboring country of Italy—and in sync with Rome's accelerated cost of living—general expenses were much lower. Vatican housekeepers (at least half of whom were males) did most of their shopping on the grounds, but it was necessary to go into Rome for such items as clothing, electrical appliances, and other durable goods. Rome supplied the Vatican with its water and its electric power, and the Vatican's own sanitation system emptied into the Roman sewers.

Citizens, who lived in assigned quarters, were not charged for electricity or telephone service, and rents were extremely low. Economic pressures and problems of a highly industrialized society did not exist within Vatican City, though salaries were rock bottom at the time: some cardinals might receive as much as $800 per month, while a monsignor might be paid a salary of $300; the commanding officer of the Swiss Guard earned about $340; and the editor of the Vatican daily paper, *L'Osservatore Romano,* got about the same amount, around $340.

There were at the time about 3,000 jobs inside the Vatican. A visitor once asked John, "Holy Father, how many people actually work in the Vatican?"

He replied jauntily, "Oh, about half of them."

The pope and members of his official "family" lived in the Apostolic Palace, a conglomeration of buildings constructed, for the most part, during the Renaissance, with some 990 flights of stairs and more than 1,400 rooms, some of which overlooked the Vatican's twenty courtyards. The palace of the Vatican was one of the world's largest, surpassed in those days only by the palace of the exiled Dalai Lama in Tibet.

The Holy Father's nineteen-room apartment on the top floor faced Saint Peter's Square. His private office, with three great recessed windows overlooking the piazza, was commodious and impressive: draped in gold damask, the windows were seldom covered by curtains; instead, whenever the sunlight beat in, the white slats on the inside shutters were closed. The papal work chamber measured 60 by 40 feet. The floor was carpeted and the walls paneled in blond wood. There were tables and satin-covered chairs spaced around the room, and books filled every inch of space in the two six-foot-high, glass-enclosed cabinets.

About five feet from the door was the pope's desk, a table with a single center drawer. On the right side of the desk, the pope kept an ornate desk clock, a high-necked desk lamp with carved statuettes at the base, a roll-blotter, and several reference books, among which were the current *Pontifici Annuario* (Pontifical Annual) and an indexed Bible. Facing the papal desk were two high-backed chairs that matched the chair on which the pontiff sat. Pope John wrote out letters and drafts of documents by hand, unlike his successor, Paul VI, who was adept with a typewriter.

On the lower floors were the apartments of the secretary of state and the master of pontifical ceremonies. The palace also housed, in one of its extensions, the Vatican museums, which contained what many experts believed to be the world's finest single collection of ancient and classical art. The museum still possesses the most important single art spectacle anywhere, the Sistine Chapel, in which the enormous *Last Judgment* of Michelangelo (restored in the early years of the twenty-first century) covers the entire wall behind the altar; the ceiling, depicting Old Testament scenes, has awed viewers for five centuries.

Beside the Apostolic Palace, the Swiss Guard had their own barracks and apartments, as they do today. Vatican City had three comparatively new apartment buildings in the early 1960s, erected to partially correct a housing shortage. There are three cemeteries within the borders of the Vatican, but they have been rarely used—with most of the world's attention focused on the vaults beneath Saint Peter's Basilica, reserved for the burial of popes.

The fenced-in Vatican Gardens were manicured year-round by a staff of twenty. There were fruit trees, cauliflower patches, plants rooted in oversize ceramic jars, and fountains of all shapes. To ensure an adequate water supply, Pope Pius XI had 9,300 irrigators installed. Fifty-five miles of pipe were laid and two reservoirs built. Each reservoir held 1.5 million gallons of water, which came directly from Lake Bracciano, outside Rome.

At John's request, the irrigation system was equipped with trick devices that squirted great jets of water at unwary visitors. When in a playful mood, the pope loved to drench new cardinals whom he inveigled to walk with him through the gardens. (The jets were dismantled by the Holy Father's successor.)

The Vatican Gardens were one of Pius's pet projects, and he frequently let the children of Vatican employees play in them. One day, noticing a school of flashy red fish swimming in one of the small ponds, he said to the youngsters who were standing nearby, "So many cardinals—and no pope!"

The next day two boys and a girl went to the pond and emptied the contents of a small pail into it. Later, when Pius went out to the garden for his stroll, he saw one extra fish in the pond. The fish was all white, like a pope.

Not far from the gardens was the "business district" of Vatican City. Located to the right of Saint Peter's Square, it could be reached by entering through the Santa Anna Gate, which was supervised by the Swiss Guard. Each visitor to the business district was asked to state the nature of his business to the guardsman on duty before he might be allowed to proceed. The roadway from the Santa Anna Gate led past the tiny parish church to the grocery store, the post office, the car pool and garage, the press office, and the offices of *L'Osservatore Romano.*

As an independent state, Vatican City had certain prerogatives with respect to Italy. For instance, in time of war, Vatican citizens and personnel were given access across Italian territory. The Vatican was exempt from customs regulations, a privilege it sometimes abused. After the end of World War II, visitors to Vatican City began picking up cartons of American cigarettes there, taking them into Italy, where American cigarettes were hard to find, and then selling them for double what they had paid. As much as this rankled officials of the Italian government (which owned a state monopoly on the sale of tobacco), nothing could be done. The practice continued even through the time of John.

The Vatican had virtually no crime. No instance of a holdup on Vatican grounds was ever recorded. Some years before John's election, however, there was one case of burglary. Only two murder attempts were recorded up through his pontificate. In one case a Swiss Guardsman, in a moment of temper, wounded his commanding officer. In the other, a demented woman shot down a priest in Saint Peter's. (Since then, in the late 1990s, there was a notorious double murder-suicide involving the Swiss Guard, a mystery that has not yet been fully resolved.)

During John's papacy, the Vatican's rarely used prison was converted into a warehouse. Most of the policemen who worked in the Vatican were laymen, as were the firemen, lawyers, stenographers, sales personnel, carpenters, bakers, gardeners, bricklayers, painters, mechanics, and other employees who kept the Vatican machinery functioning. To supplement this lay staff, a number of small religious societies provided services of various types. For instance, the Vatican telephone system and local mail deliveries were handled by the friars of the Little Work of Divine Providence. A group of nuns, affectionately known as the Sisters of Tapestry, specialized in the mending and restoration of the thousands of precious tapestries that adorn the walls of the Apostolic Palace. The Do-good Brothers operated the Vatican pharmacy, and on a nearby island in the Tiber River, they administered a hospital, where during the Nazi occupation of Rome they earned a reputation for hiding Jewish refugees and American and British pilots shot down in combat.

Another religious group, the Salesians of Saint John Bosco, provided the Vatican with typesetters and linotype operators. Charged with printing secret and confidential Vatican documents, the members of this group also ran a printing plant, which published documents in 120 different alphabets and languages, including hieroglyphics, Chinese, Braille, Hebrew, Arabic, and Coptic.

Perhaps the most unusual job in the Vatican in that era was performed in a high-ceilinged room in the Apostolic Palace. The room was lined with shelves and drawers containing ashes, slivers of bones, and other remains of early saints and martyrs. Under an electric lamp in one corner of this strange chamber, the world's most macabre library, a Vatican officer worked, surrounded with tiny boxes and envelopes addressed to all parts of the globe. These were for the purpose of conveying saintly relics. According to canon law in those days, a relic had to be enclosed in every altar of every church. Because new churches were opened all the time, authentic relics were in constant demand. The librarian was kept continually busy.

Most jobs were quite ordinary: the pontifical shoemaker, for example. Since 1939, the task of making papal shoes had belonged to Telesforo Carboni, who referred to John as "a wide 10." Like many other shoemakers, Carboni was quite a raconteur, particularly on the matter of footwear.

> I remember the time Pope John, who had a big foot, which could take even a ten and a half, came to me and said, "Signor Carboni, you must make me a pair of shoes that are nice and big and don't cramp my feet."
>
> A man with cramped feet, you know, will usually have cramped ideas in his head, and so His Holiness wanted a pair of shoes that wouldn't cramp him in his work.
>
> The pope didn't have corns on his feet, but he did have a high instep, and the top of a shoe, if it was a bad fit, could cut his foot when he walked. He showed me the most comfortable pair of shoes he had ever had, made by his nephew, a shoemaker in Bergamo, and they were died purple. I was horrified at the color. Who ever heard of a pope wearing purple shoes?
>
> "Holy Father," I said, "you can't wear purple shoes. It is not the pope's color."
>
> Pope John thought for a bit, then he said, "But, Signor Carboni, I don't want to hurt my nephew's feelings. When I write him, I must tell him I am wearing the shoes he made for me."
>
> "*Ci penso io,*" I said. (I'll take care of it.)
>
> "*Benissimo!*" (Great!), exclaimed His Holiness. "You have solved my problem. You are a saint. You have made the first miracle of my reign!"

In John's era, as it had been for popes since the fifth century, the office of the papacy held a unique place in the public imagination. Just a listing of his titles boggles the mind: Bishop of Rome, Successor of the Prince of the Apostles, Supreme Pontiff of the Universal Church, Servant of the Servants of God, Patriarch of the West, Primate of Italy, Archbishop and Metropolitan of the Roman Province, and Sovereign of the State of Vatican City. Each title signifies a historical development in the papal office and reflects an element in the fabric of responsibilities in which he is invested.

When elected, the pontiff loses the civil ties that have bound him to the nation of his origin. He finds that his daily life is governed, down to the most minute detail, by long-practiced tradition. In John's time, the pope's confessor, an ordinary priest, had to be a Jesuit, and he was required to visit the Vatican once a week at a fixed time—and he alone could absolve the pope of his sin. The master of the Apostolic Palace had to be a Dominican, the sacristan an Augustinian. It would be exceedingly difficult to change any of these protocols, lest the religious congregation in question regard the action as an affront of some kind against their number.

The first time John received his relatives in a special audience, shortly after his coronation, the visitors approached the new pontiff timidly, and when they saw him vested in his pontifical white robes, they knelt and bowed their heads.

"Lasciate perdere!" (Forget all that!), John said. "Don't be afraid. It's only me!"

Pope John was the first among equals of all the bishops in the world, all of whom came under his direct authority. He possessed, in theory, full and absolute power over the Roman Catholic Church. Any decree issued by the Holy See required his signature. He could obey or ignore precedent. He could set aside tradition and write—or completely rewrite—constitutions of the

Church, and he could change discipline (such as the requirement of priestly celibacy) without consultation. The pope was empowered to proclaim dogmas on his own authority, though on matters that touched upon the life of the Universal Church it was customary (and still is) for the pontiff to consult with the bishops as the magisterium, or teaching authority, of the institution. In fact, the last time a pope defined a new doctrine of the Church infallibly and *ex cathedra* (from the chair) was in 1950, when Pius XII proclaimed the dogma of the Assumption of the Blessed Virgin Mary.

John, like most modern popes—perhaps more than most— viewed himself a temporary occupant of an eternal office and maintained a sense of irony and humor about his position in the Church and in the world, something he had learned over his long diplomatic service and the surprises and near-humiliations that he often encountered during those years.

The pope, he knew, could be judged by no man, and there was no appeal from his decisions. In this respect his position was that of a head of state who could not be brought to court. Acting in his executive authority, John had the exclusive power to approve or sanction or suppress religious orders; grant indulgences to sinners; beatify or canonize saints; appoint bishops (perhaps the single most far-reaching power in his arsenal); erect, administer, or suppress dioceses; assign an auxiliary or co-adjutor bishop in a diocese; found and legislate for pontifical universities; publish liturgical and theological books; administer the temporal goods (including cash) of ecclesiastical foundations; and erect and govern overseas missions dependent on the Holy See.

As a legislator, John could convoke and preside over ecumenical councils, regulate holy days and Catholic feasts, introduce new rites and abolish old ones, issue *ex cathedra* decrees on matters of faith and morals, introduce or suppress Church laws on any subject, defend doctrine against heresies, and define feast days and periods of fasting throughout the whole Church.

Then, as chief judge, the Holy Father was juridically empowered, among other things, to rescind vows and oaths for members of religious congregations who wished to return to secular life, give matrimonial dispensations, act as a court, establish rules of judicial procedure, establish censures or punishments for crimes, organize courts for hearing cases, and organize courts or appoint synodal judges for the Diocese of Rome.

In those years there were no provisions for the possible incapacitation of a pope. He was elected for life and, often a septuagenarian or an octogenarian, would serve until his death. Although a very few popes in the past had abdicated, there was little or no thought during John's reign that it could ever come to such a decision in the modern age. Even the College of Cardinals was powerless in the face of an ill or incapacitated pope.

As head of the Holy Roman Church, the pope ran a vast business organized as a corporation, directing twelve congregations (also called dicasteries) of cardinals—known as the Roman Curia or "court"—a system that dated from the late sixteenth century, during the reforming period of the Council of Trent.

Members of the Roman Curia had warned John that the ecumenical council would take years to prepare, that organizing it by 1963—as he had hoped—would be impossible. To which he replied, "Good, then we'll have it in 1962." In May 1959, he established the Pre-Preparatory Commission, headed by Cardinal Tardini, whose original plan was to send a questionnaire to everyone eligible to vote at the council—that is, 2,598 bishops, as well as heads of (male) religious orders and heads of thirty-seven Catholic universities. Instead, however, John and Tardini decided on a letter, which would not lock recipients in to a certain type of answer, the way a questionnaire might. The letter read, in part:

> The venerable Pontiff wants to know the opinions or views and to obtain the suggestions and wishes of their excellencies, the bishops and prelates who are summoned by law (Canon 223) to take part in the ecumenical council. These [suggestions] will be most useful in preparing the topics to be discussed at the council.

Of the 2,598 letters the Vatican sent out, they received 1,998 letters in return, a 77 percent response rate that would be the envy of any direct marketer. The responses came from around the world, in letters that ranged from six lines written by a bishop in Australia to twenty-seven pages penned by a bishop in Mexico. They were eventually published after the council in eight volumes totaling 5,000 pages. The Curia were invited to add their ideas, which brought in an additional 400 pages.

At the time, however, the material was confidential. It filled 2,000 massive folders. To the relief of certain members of the Curia, the bishops of the world seemed not, at this stage, to have fully grasped the opportunity being provided them. Many of the responses to the pope's letter dealt with concerns about the spread of communism or a desire on the part of bishops to see more faculties provided for their dioceses. A few people wanted to discuss furthering the role of the laity in the Mass (as well as using the vernacular instead of Latin) and a few more wanted to discuss the controversial question of priestly celibacy. The pope, however, was a firm supporter of celibacy.

Reading and organizing all this material took a good deal of time. In May 1960, Pope John announced the next step toward the council, the formation of ten preparatory commissions, whose job it was to catalogue each letter. The idea was that the bishops would then be able to examine these documents and come to the council prepared to vote and discuss specific issues. Each commission was led by the head of the corresponding congregation within the Curia. Cardinal Gaetano Cicogani, for instance, head of the Congregation of Rites, would be responsible for the council's document on liturgy. And Cardinal Alfredo Ottaviani, head of the Congregation of the Holy Office, formerly the Office of the Inquisition, would rule over the Theological Commission.

Ottaviani's influence would extend far beyond that one commission, however. Blind in one eye, Ottaviani could, as the saying went around the Vatican, see more with one eye than most could with two. The motto on his Vatican coat of arms was *Semper idem* (Always the same), fittingly enough, since he was the leader of the conservatives who inhabited the Curia. On every issue, he argued forcefully that the Church should at all times resist change. Theology, he said, was for the protection of the truths that already existed—not for the promulgation of new ideas.

It was understandable that John had organized his preparatory commissions to echo the structure of the Curia in order to take advantage of organizational machinery that already existed. John also may have understood that excluding the Curia from a process that might naturally fall within their purview might alienate them even further from the idea of an ecumenical council. But liberals within and outside the Church were concerned that the Curia controlled so much of the agenda for the council. This concern extended to the 800 theologians who were members of the preparatory commissions and who actually created the documents; most of these men were not viewed as forward thinking, to say the least.

Finally, liberals were dismayed at what later occurred at the Roman Synod in January 1960, which many viewed as a kind of practice run for the upcoming Vatican Council. The synod (a Greek word for "assembly," used especially in regard to Church meetings) was held to address specific issues in the Diocese of Rome, but it mainly rubber-stamped previous and long-held Curia decisions regarding priests—for instance, that they were forbidden from driving cars unless in cases of emergency, and they could not be alone with a woman, a communist, or a heretic.

Not exactly illuminating ecumenical fodder. But John had a trick left up his sleeve. In addition to creating the preparatory commissions, he created a so-called Secretariat for Promoting Christian Unity and named as its head Augustin Bea, a Jesuit he had promoted to cardinal on December 14, 1959. As Ottaviani was to the conservatives, Bea was to the liberals. Eighty years old, so frail, as one historian has written, "that it seemed a puff of wind would blow him over," Bea had been around the Curia for a long time. He had been Pius XII's personal confessor (and so could not be attacked on grounds that he was against the previous Vatican regime) and was tough enough to survive and triumph through the worst curial infighting. A secretariat did not quite have the status of a commission, but John made sure everyone knew how important he considered the position. When he named Bea to the position, he articulated its importance: "In order to show in a special manner our love and good will towards those who bear the name of Christ, but are separated from this Apostolic See."

This was a signal to Christians of all faiths that the pope was seriously interested in beginning the process of ecumenical reunion. As Bea himself put it, "So we may say . . . that the council should make an indirect contribution to union, breaking the ground in a long-term policy for preparation for unity." In terms of theology, Bea (along with the pope) felt that while the so-called revealed truth of God was immutable, its formulation—how it was expressed and delivered to the faithful—was not. The key word for both John and Bea was *aggiornamento*—"updating." The council would be all about expressing the old in a new way.

CHAPTER TEN

World in Crisis
(January 1961–September 1962)

P ope John XXIII promulgated another encyclical, *Mater et magistra* (Mother and Teacher), on May 15, 1961, on the subject of Christianity and social progress. It was issued on the seventieth anniversary of Pope Leo XIII's *Rerum novarum* (On the New Things), which has defined Catholic social teaching for more than a century.

John's 25,000-word encyclical starts off after its customary greeting with "Mother and Teacher of all nations—such is the Catholic Church . . ." and invokes the Church's "maternal care" toward both individuals and nations.

Although this did not garner notable headlines at the time, it is part of a continuum of John's reaching out to the broader world. He does so through the gospel, at the very outset noting that although Christ was foremost concerned with the eternal salvation of people, "He showed His concern for the material welfare of His people when, seeing the hungry crowd of His followers, He was moved to exclaim: 'I have compassion on the multitude.' And these were no empty words. . . . He proved them by his actions, as when He miraculously multiplied bread to alleviate the hunger of the crowds" (para. 4). As if to underscore "these were no empty words," the entire remainder of the encyclical serves as a mandate and a how-to manual, as it were, for fulfilling Christ's compassion on the multitude as told in Mark 8:2.

A significant portion of the encyclical reaffirms prior teaching, especially of Leo XII in *Rerum novarum,* but also of Pius XI in *Quadragesimo anno* (In the Fortieth Year) and Pius XII in a radio address. *Mater et magistra* ultimately strikes out on its own in tone and substance. It reaffirms the principle of subsidiarity and gives a blessing to socialization, a term widely misinterpreted. Regarding socialization, Pope John was "aware that the mindset of the modern state is toward what he calls 'socialization'—'the fruit and expression of a natural tendency, almost irrepressible in human beings, the tendency to join together to attain objectives which are beyond the capacity and means at the disposal of single individuals.' But socialization does not necessarily turn men into automatons. 'For socialization is not to be considered as a product of natural forces working in a deterministic way. It is, on the contrary, as we have observed, a creation of men, beings conscious, free and intended by nature to work in a responsible way."

In restating elements of *Rerum novarum,* John recognizes the right of workers to associate, as in unions. And he declares two extremes as being contrary to Christian principles and the nature of man: "unrestricted competition in the liberal sense, and the Marxist creed of class warfare." Citing Pius XI, he restated oppo-

sition even to what was termed "moderate Socialism," criticizing
it for a sole focus on material well-being and production. Describ-
ing the Great Depression, he tells how capitalistic

> unregulated competition had succumbed to its own inherent
> tendencies to the point of practically destroying itself. It had
> given rise to a great accumulation of wealth, and, in the pro-
> cess, concentrated a despotic economic power in the hands of
> a few [to quote Pius XI] "who for the most part are not the
> owners, but only the trustees and directors of invested funds,
> which they administer at their own good pleasure."

John was also prescient and attentive to the growing importance of emerging nations in Asia and Africa, specifically mentioning them as they broke from colonial rulers. As with *Pacem
in terris,* he underscored the growing importance of economic in-
terdependence. He saw a global village well before it was a commonplace.

In giving his purpose for the encyclical, the pope notes he is
not merely commemorating Leo XIII or his successors. His job
is to clarify and build on their work. Right away, he endorses
the "principle of subsidiary function," which Pius XI laid out in
Quadragesimo anno. In plain terms, the principle sets up a pyramid whereby the individual is the first recourse in private en-
terprise, scaling upward to communities and governments only
after these first options are tried. Yet John does not stop there.
He expressed "unbearable sadness" over mass unemployment and
"subhuman conditions" and urgently called to redress terrible in-
equities within nations and between nations.

In *Economic Justice,* John Pawlikowski, O.S.M., and Donald
Senior, C.P., wrote:

Pope John was severely criticized by right wing authors because he understood the principle of subsidiarity as a basis for government intervention on behalf of those who need help. Nevertheless this was a legitimate development. Pope John did not take away initiative or creativity from persons or small groups. He simply recognized that subsidiarity means help and help is often badly needed.

It might be mere semantics and revisionism to ask how John himself might comment on the contemporary issue of so-called same-sex marriage. We know he did "solemnly proclaim that human life is transmitted by means of the family, and the family is based upon a marriage which is one and indissoluble and, with respect to Christians, raised to the dignity of a sacrament." But it is more intriguing to wonder how he might apply these words to modern mores, including artificial insemination and other fertility measures: "The transmission of human life is the result of a personal and conscious act, and, as such, is subject to the all-holy, inviolable and immutable laws of God, which no man may ignore or disobey. He is not therefore permitted to use certain ways and means which are allowable in the propagation of plant and animal life."

We might even wonder whether the Good Pope would today be weighing in on such matters as genetically modified organisms (GMOs). What we do know is that *Mater et magistra* spends a great deal of time talking about agriculture, rural development, family farms, price protection, credit banks, social insurance, taxation, and economic structures affecting farming. In fact, in a sense no other topic arguably receives as much verbal volume as agricultural issues. The pope not only saw the critical connection of food production to world economic conditions, but also, possibly because of his own country roots, saw firsthand the human values inherent in working the land.

As John tackled the issue of population growth, he rejects it as an opportunity for artificial birth control. In what some might consider a too-sunny view of science and technology, the pope finds optimism in the "almost limitless horizons" afforded by science and technology in service to providing the necessities of life.

Yet again, though, he scolds nations for their lack of solidarity or their deficient economic systems and organizations. "Attention must then be turned," he wrote, "to the need for worldwide cooperation among men, with a view to a fruitful and well-regulated interchange of useful knowledge, capital and manpower." Without understanding the coded references to existing political systems, one might argue that the proposed solutions lack sufficient clout and specificity. Is the pope arguing that essentially the world has enough goods, including food? That it is merely a matter of equal and fair distribution? Perhaps experts would see validity in that, and the debate surely continues.

However, John naturally sees—and faces squarely—the abhorrent dangers wrought by technology, especially nuclear weapons. "We are sick at heart, therefore, when we observe the contradiction which has beguiled so much modern thinking. On the one hand we are shown the fearful specter of want and misery which threatens to extinguish human life, and on the other hand we find scientific discoveries, technical inventions and economic resources being used to provide terrible instruments of ruin and death."

Although critics and supporters alike make much of *Mater et magistra*'s positions on issues of economics and social justice, one hears very little about making Sunday holy. The pope devotes six paragraphs (248–53, inclusive), a significant segment, to this topic. It rarely, if ever, gets mentioned in any contemporary discussion.

He frames the discussion in terms of ancient biblical prescriptions (the Third Commandment) pitted against the secular world, in the context of labor and society and human dignity:

> Free from mundane cares, he [man] should lift up his mind to the things of heaven, and look into the depths of his conscience, to see how he stands with God in respect of those necessary and inviolable relationships which must exist between the creature and his Creator.
>
> In addition, man has a right to rest a while from work, and indeed a need to do so if he is to renew his bodily strength and to refresh his spirit by suitable recreation. He has also to think of his family, the unity of which depends so much on frequent contact and the peaceful living together of all its members.
>
> Heavy in heart, We cannot but deplore the growing tendency in certain quarters to disregard this sacred law, if not to reject it outright. This attitude must inevitably impair the bodily and spiritual health of the workers, whose welfare We have so much at heart.
>
> In the name of God, therefore, and for the sake of the material and spiritual interests of men, We call upon all, public authorities, employers and workers, to observe the precepts of God and His Church and to remember their grave responsibilities before God and society.

The pope's words are worth pondering. How does the Christian carve out this sacred time in an increasingly secular world? Is it reasonable to expect Christians to take the Sabbath as seriously as brothers and sisters in other religions? What are the responsibilities of employers in this regard? And ironically, though the passage encourages recreation, should Christian parents whose children are forced to participate in sports protest when it conflicts with worship times? Such strong words from a pontiff about making Sunday holy may seem archaic to many ears in the twenty-first century. But there is no denying that John saw it as an important issue, worthy of serious attention—attention that has ostensibly been missing for more than fifty years.

Mater et magistra altered the course of Catholic social thought, not only opening the way for future dialogue with other ideologies but also setting the tone for conciliatory discussions—and even serving as a precursor for topics that would be addressed in Vatican Council II.

In the U.S. Catholic Church, *Mater et magistra* arguably marked a line in the sand in the debate over the Church's social justice positions. In a foreshadowing of the acrimonious tone that would mark political discourse in the United States some fifty years later, the Jesuit magazine *America* and the conservative flagship *National Review* jousted publicly over the import of John's *Mater et magistra*—a far cry from the almost universal acceptance that would greet *Pacem in terris* two years later. Christian liberals and conservatives still cite this 1961 encyclical as they stake out positions on the role of government and seek intellectual fodder for their case. Most would likely agree with this early assessment of the encyclical by the *Economist:* "The encyclical represents a shift to the left in the Church's attitude." By opening the door to more government action for the common good and taking a less strident approach to doctrines opposed by the Church, the encyclical was seen as a softening of some traditional views espoused by Leo XIII and succeeding popes. Although the Church continued to oppose naked capitalism (economic "liberalism" in the words of the pope), critics like William F. Buckley Jr. were suspicious, to say the least, of the shift in emphasis. In the July 29, 1961, issue of his *National Review,* Buckley wrote, "Whatever its final effect, it must strike many as a venture in triviality coming at this particular time in history." He lamented that "insufficient notice is taken (of) the extraordinary material well-being" of countries like Japan, West Germany, and the United States and complained that little attention was paid to the economic failures of communist nations.

A few months later, in the August 12 issue of *National Review,* a small item read, "Going the rounds in Catholic conservative circles: 'Mater si, Magistra no.'" This phrase was later revealed to have been jokingly coined by columnist Garry Wills in a phone conversation with Buckley. (The phrase was especially volatile because rallies by Fidel Castro featured chants of "Cuba si, Yankee no.") A salvo was returned by *America* magazine. "It takes an appalling amount of self-assurance for a Catholic writer to brush off an encyclical of John XXIII. . . . Mr. Buckley was equal to the challenge. It takes a daring young man to characterize a papal document as 'a venture in triviality.'" Later Buckley explained his stance:

The editorial in question spoke not one word of criticism of the intrinsic merit of *Mater et magistra.* Our disappointment was confined to the matter of emphasis, and timing, and by implication, to the document's exploitability by the enemies of Christendom, a premonition rapidly confirmed by the encyclical's obscene cooption by such declared enemies of the spiritual order as the *New Statesman* and the *Manchester Guardian,* which hailed the conversion of the pope to Socialism!

Later in life Buckley conceded he would have preferred the heading had not found its way in print.

Tit for tat continued. Philip S. Land, S.J., in *America:* "As one whose life work has been the assimilation and the teaching of Catholic social doctrine, I can say with absolute certainty that no collaborators Pope John might have turned to—European or American or other—could have or would have prepared an encyclical that would be acceptable to the editors of the *National Review.*"

Commentaries in the mass media were typically more amicable, and more positive.

TIME magazine extensively quoted Protestant theologian Reinhold Niebuhr's essay in the *Christian Century:* "From the standpoint of the *Mater et magistra* encyclical, what could be clearer than that the path from the Thomistic theory of a just price based upon labor value, to the theory of Adam Smith, guaranteeing social justice by the automatic balances of a free market, descends steeply from the heights of justice to the morass of private greed?"

Pope John's encyclical ignores its own indebtedness to some of the moral achievements of the welfare state and foreign aid, says Niebuhr, but "before we ungenerously attribute to conscious and unconscious cribbing from a culture it ostensibly abhors the massive achievement of modern Catholicism in adjusting to the realities of modern industrialism," it is necessary to recognize that Catholicism has traditions that make this adjustment possible.

The Roman Church, according to Niebuhr, balances concern for the individual with concern for the health of the community, which is to be achieved by what the encyclical calls "objective justice and its driving force, love." Says he, "To assert that justice is the norm and 'love the driving force' is certainly a theory of the relation of . . . love to the social order preferable to some Protestant and secular theories."

Niebuhr, who has long lashed out against the perfectionist strain in Protestantism, further admires the Roman Catholic Church for having relegated its perfectionists and ascetics to the monasteries, where they cannot mess up the proper processes of society, full of contingencies and compromises.

John's encyclical displays "dated rather than eternal wisdom," Niebuhr believes, in opposing birth control and ignoring the fast pace of population increase. But he refrains from laboring the point, "lest the professional anti-Catholics take too much courage. They regard the Roman church as a monster. It is really a very impressive survival from medievalism, which has managed to apply its ancient wisdom to the comfort of a harassed generation in a nuclear and technical age."

John had "grown in confidence and sureness of touch," according to biographer Peter Hebblethwaite, and with the publication and reception of *Mater et magistra,* he had found his voice as the supreme pontiff and teaching pastor of the whole Church.

He issued *Aeterna Dei sapientia* (God's Eternal Wisdom), on November 11, 1961. From May 1961, when *Mater et magistra* came out, to November 1961, the world had changed. Cold War tensions had increased in August, when the East Germans built the Berlin Wall. The move further isolated countries behind the Iron Curtain and increased the already grave concerns for the Church in Soviet satellite countries such as Poland, Hungary, and Czechoslovakia.

Although some accounts at the time of the encyclical's issuance zeroed in on Cold War tensions, the heart of the encyclical lies elsewhere. As for the plight of the Church in embattled areas, the pope invokes the spirit of Leo I the Great (440–461):

> The same waves of bitter hostility break upon her [as they did in Leo's time]. How many violent storms does she not enter in these days of ours—storms which trouble our fatherly heart, even though our Divine Redeemer clearly forewarned us of them!
>
> On every side we see "the faith of the Gospel" imperiled. In some quarters an attempt is being made—usually to no avail—to induce bishops, priests and faithful to withdraw their allegiance from this See of Rome, the stronghold of Catholic unity.
>
> To those of you who suffer patiently in the cause of truth and justice, we speak the consoling words which Saint Leo once addressed to the clergy, public officials and people of Constantinople: "Be steadfast, therefore, in the spirit of Catholic truth, and receive apostolic exhortation through our ministry."

Aeterna Dei sapientia's stated goal, as reflected in its published title with heading, was to commemorate 1,500 years to the day since the death of Pope Saint Leo I, in 461, and to shine a light on the "See of Peter as the center of Christian unity." The 6,500-word encyclical, the sixth of John's pontificate, on the surface celebrates the life and death of Leo the Great—"the greatest among the great," as Pius XII dubbed him—but it was also a vehicle to promote Roman Catholic orthodoxy, the power of the Apostolic See, and to revisit ancient schisms. All this was with an eye toward the Vatican Council that would start within less than a year. (At this juncture, an opening date for the council had yet to be announced.)

Before exploring the broader purposes and impacts of the letter, a summary of its contents is in order.

John spends significant portions of his document reviewing the spiritual biography of Leo I. He presents not so much a chronological, linear biography as a blow-by-blow account of the victories of this great defender and doctor of the Catholic Church. As recorded in history books, no episode in Leo's life looms larger than his encounter with Attila the Hun, a turning point in the history of Western civilization:

> Wherein, then, lies the true greatness of this pope? In moral
> courage?—in that moral courage which he showed when,
> at the River Mincius in 452, with no other armor to protect
> him than his high-priestly majesty, he boldly confronted
> the barbarous king of the Huns, Attila, and persuaded him
> to retreat with his armies across the Danube? That was
> certainly an heroic act and one which accorded well with the
> Roman pontificate's mission of peace.

Then the pontiff considers Saint Leo through three perspectives: as a faithful servant of the Apostolic See, as the Vicar of Christ on Earth, and as a doctor of the Church.

Regarding the first characteristic, servant of the Apostolic See, John recalls how Leo soared from deacon to pope, in 440, on the strength of his theological and diplomatic acumen, with John perhaps identifying with the latter set of skills based on his own diplomatic experience. Dying in 461, Leo was one of the longest-reigning popes—a contrast John surely recognized as he neared his eightieth birthday at the time of the encyclical's publication.

As for Leo's tenure as Vicar of Christ, John exults that "rarely in her history has Christ's Church won such victories over her foes as in the pontificate of Leo the Great. He shone in the middle of the fifth century like a brilliant star in the Christian firmament." Leo claimed decisive victories over heretics throughout his papacy. In this regard, John compared Leo to the great Saint Augustine of Hippo and Saint Cyril of Alexandria. Augustine famously attacked the Pelagians, who asserted that salvation could be attained by one's own will, and, as John wrote, Augustine "insisted on the absolute necessity of divine grace for right living and the attainment of eternal salvation." Likewise, John pointed out, "Saint Cyril, faced with the errors of Nestorius, upheld Christ's Divinity and the fact that the Virgin Mary is truly the Mother of God."

Upholding Leo as doctor of the Church's unity, John inventories Leo's battles defending the Church's dogma, most notably on the incarnation. These examples of Leo's strong leadership included his condemnation of the Ephesus Council in 449 because it held that Christ had only a divine nature, branding it a "robber council." Two years later, Leo summoned the Council of Chalcedon, but only on his terms and not the emperor Marcian's terms, marking a strong assertion of the pope's magisterial authority.

John also highlighted Leo's sermons and epistles, and his devotion to truth, harmony, and peace.

At this point, it's not hard to draw parallels with John and his upcoming council. John himself proceeds to build his case for the upcoming Second Vatican Council, standing on the shoulders of Saint Leo, as it were.

Venerable Brethren, the time is drawing near for the Second General Council of the Vatican. Surrounding the Roman Pontiff and in close communion with him, you, the bishops, will present to the world a wonderful spectacle of Catho-

lic unity. Meanwhile We, for Our part, will seek to give
instruction and comfort by briefly recalling to mind Saint
Leo's high ideals regarding the Church's unity. Our intention in so doing is indeed to honor the memory of a most
wise pope, but at the same time to give the faithful profit-
able food for thought on the eve of this great event.

Was John trying to get in front of his possible detractors in the
Curia? Or was he perhaps giving himself strength and encouragement by seeing Leo as an exemplar of papal power? Are his
words even partly an exercise in wishful thinking, positive thinking that he hopes will translate to positive outcomes?

John spends significant parts of his encyclical invoking Saint
Leo's vigorous defense of the pope as the Bishop of Rome, of the
necessity of Rome as the center of unity, going so far as to describe
it as a spiritual haven. He also steadfastly reasserts Leo's defense
of the supreme authority of Peter and his successors.

Of course, none of that was likely to offend Catholics, but
what of those separated Christian brethren that the council and
the pope himself were reaching out to? How would this sound
to their ears? As *TIME* magazine put it, "*Aeterna Dei sapientia*
forcefully reflected the pope's own oft-expressed dream of healing
the breach between Christendom's largest branches. But to many
Protestants and Orthodox Christians, the encyclical seemed as
much a reminder of unacceptable papal claims as a warm appeal
for unity."

It's as if John was tightrope-walking. He satisfied Church conservatives, and long-standing Church teaching, by aligning with
Leo's endorsements of papal power. At the same time, however,
he ardently wanted to reach out to his separated brothers. This
dual approach is evident in several key passages and his own ac-

tions. That same November, he met with Arthur Lichtenberger, bishop of the Episcopal Church, the first time an Episcopal bishop met with the pope at the Vatican. In other words, he might have toed the line doctrinally, but he was always welcoming personally.

Nor did the great honors paid to Leo by the official representatives of the Eastern churches terminate with his death. The Byzantine liturgy keeps February 18 as his feast day, and most truly proclaim him as "leader of orthodoxy."

> Our purpose, Venerable Brethren, in focusing attention on these facts has been to establish beyond doubt that in ancient times East and West alike were united in the generosity of their tribute to the holiness of Saint Leo the Great. Would that it were so today; that those who are separated from the Church of Rome yet still have the welfare of the Church at heart, might bear witness once more to that ancient, universal esteem for Saint Leo.
>
> We are fully confident that this solemn assembly of the Catholic hierarchy [the council] will not only reinforce that unity in faith, worship and discipline which is a distinguishing mark of Christ's true Church, but will also attract the gaze of the great majority of Christians of every denomination, and induce them to gather around "the great Pastor of the sheep" who entrusted His flock to the unfailing guardianship of Peter and his successors.

Finally, John ends on a curiously militant note. The tone makes sense when it comes to opposing totalitarian regimes, but it is potentially troublesome when viewed from a Protestant or Orthodox vantage point, since members of these churches were not apt to rally around the standard that is Rome:

We cannot end this encyclical, Venerable Brethren, without referring once more to our own and Saint Leo's most ardent longing: to see the whole company of the redeemed in Jesus Christ's precious blood reunited around the single standard of the militant Church. Then let the battle commence in earnest, as we strive with might and main to resist the adversary's assaults who in so many parts of the world is threatening to annihilate our Christian faith.

At Christmas 1961, the world stage was fraught with tense players. Algerian nationalists were talking peace in their war with the French. Cuba's tilt toward communism was rattling America's nerves. Taiwan's Quemoy and Matsu islands were flashpoints "almost within the jaws of Communist China." And the Berlin Crisis was fresh and volatile, with much made of Cardinal Spellman's walking six feet into East Berlin territory for a Christmas visit.

But there were signs of hope, too. Amid these Cold War anxieties, pilgrims to Bethlehem encountered "the most friendly atmosphere . . . since before the Palestine war of 1948 cut Jerusalem in two." Soviet Premier Nikita Khrushchev conceded that "neither he nor the Communist party is infallible."

Against this backdrop, on December 25, 1961, John promulgated his apostolic constitution *Humanae salutis* (Of Human Salvation). On one level, the document is formal and technical. It officially convokes the ecumenical council whose planning was well under way and sets 1962 as the date for its start (nothing more specific than that; the actual start date of October 11, 1962, was proclaimed in a motu proprio less than two months later, on February 2). This made official what had been stated in January 1959, when the pope announced his plans for the worldwide assembly. But on a broader scale, the declaration delineated the pope's intentions for the council and clarified the themes John held dear. In more subtle terms, for those who delighted in reading the Vatican's tea leaves, one could search the pope's words for "coded" messages to the Curia as well as to other Christian churches and to the world at large.

Church prelates read the pronouncement in Rome's four basilicas, first at Saint Peter's and then at Saint Paul's Outside the Walls, Saint John Lateran, and Saint Mary Major.

Seen as one pronouncement sandwiched among others, *Humanae salutis* contained more messages of optimism. John's Christmas message in these same days referred to encouraging signs in a tense world and cited his own recent *Mater et magistra* as a basis for hope and peace. Although the official promulgation of the council came amid these hope-filled Christmas tidings, the document is more than a dose of holiday good cheer. It also provided John with an opportunity not only to outline the agenda for the upcoming council but to advance his influence over stalwart and starchy Church insiders who would just as readily have stalled any progress toward the council.

Humanae salutis, in a little more than 3,000 words and broken down into twenty-four sections, put Christ at its center and in the center of the secular world. "The Redeemer of man's salvation, Christ Jesus" is not only its rhetorical opening but also its heart. In the very first paragraph, John quoted Matthew's Gospel, "Here I am with you always, until the end of the world," and this assurance infused the whole message—and John's hope and prayer for the council. John's second scriptural reference, from the Gospel of John in the second paragraph, reflected his own brimming optimism: "Be of good cheer, I have overcome the world!"

Some of the pope's rationale for the council was standard fare for papal documents. He lamented the world's lack of spiritual progress and its rejection of God and the Church. But he went a step further than usual, calling for a "complete renovation" of the human community. As for confronting the world, he saw the Church's role as nothing less than "to intervene actively in all spheres of human activity."

John spoke plainly: the council's job was to renew the world by first renewing the Church. To use his analogy, a small seed was to grow into a large tree. John saw this renewal as the path to peace and unity.

> Therefore we are confident that . . . an ecumenical council will . . . not only light up a fervent Christian wisdom and fortify the inner energy of the soul, but also *pervade the whole of human activity.* [emphasis added]
>
> We gave you the first announcement of the celebration of the Ecumenical Council 25 January 1959. By doing this, it seemed that with heart and trembling hands, threw a sort of small seed. . . . From that day I spent almost three years, during which we saw that tiny seed grow under the breath of heavenly grace into a giant tree.

But *Humanae salutis* also bore the conciliatory imprint of John's personality that would later be demonstrated most cogently in his master work, *Pacem in terris* (Peace on Earth). The last several sections were sprinkled with important all-embracing phrases: "Christian" instead of "Catholic"; "Christians" instead of "separated brethren." Its tone seemed more diplomatic and welcoming than that displayed in *Aeterna Dei sapientia,* issued only a month before. Other examples abounded, with emphasis added: "We ask for all the faithful and *all Christian people* to devote full attention to address the council and fervent prayers to Almighty God. . . ." "With these our exhortations intend to consult with the *beloved children of both clergy of all nationalities and all Christians in any category.*"

Chapter Eleven

Aggiornamento, Si! (October 1962)

On July 15, 1962, the Vatican sent out 2,850 invitations to those who would deliberate at the council, which was set to open on October 11 at Saint Peter's Basilica. These invitees included eighty-five cardinals and 2,131 bishops, as well as abbots and the superior-generals of religious orders of men. They came from all over the world. Also invited were *periti*, or theological experts, there to help the bishops.

Pope John XXIII made sure to invite observers-delegates from other religions. A representative from almost every non-Catholic religion attended the council, and each one was given a good seat. And, because the business of the council was conducted in Latin, the Vatican also provided translation services. Each observer-delegate was also invited to a smaller, more private meeting as honored guests "in the house of their father," as John said.

As October approached, Rome buzzed with the excitement of visitors—council attendees, tourists, and, because of the magnitude of the event, the world's press. About 1,200 reporters from around the world covered the opening of the council, including one young unaccredited American journalist who disguised himself in a cassock and managed to sneak through three levels of security, including Swiss Guardsmen and Vatican plainclothes detectives.

Saint Peter's itself was spruced up with new bathrooms, two coffee bars, thirty-seven microphones—one observer said that "a whispered note can be heard in the remotest part" of the Church—and 2,900 seats. This alone cost the Vatican $1,000,000 (about $7,000,000 in today's dollars). This did not include transportation and lodgings for those bishops who could not afford to pay their own way, all of which would severely tax Vatican coffers during the next four years.

This extraordinary council would be public, for all the world to see. There was the feeling that, unlike previous councils—which were directed mainly to clergymen—decisions made here would affect Catholics as they went about their everyday worship. In a sense, the still-unstated aims of the council—unstated in part because John had deliberately not done so—indicated that the council itself would be one of broad focus and scope. Anything could happen.

On October 4, 1962, just one week before the council opened, John traveled by train to the tomb of Saint Francis of Assisi for his feast day and then on to Loreto, to the shrine of the Holy Virgin. The distance was only a couple of hundred miles, but it was the first time that a pope had traveled by train in a century, since all popes during that time were more or less "prisoners of the Vatican." People flocked to stations along the way, waving and cheering. Fifty thousand gathered at Loreto, where John prayed at the shrine of the Blessed Virgin for the success of the council. "This date in my life should be written in gold," he told the well-wishers.

Only a few people knew that John was seriously ill. In the week leading up to September 23, he had undergone a battery of tests, which determined that certain intestinal pains he was experiencing indicated stomach cancer, probably of the same type that had afflicted others in his family. There was no cure, and his secretary, Monsignor Loris Capovilla, wrote of the "unexpected and disconcerting news of the illness that threatened the life of the pope." It was decided that for the time being this startling news should be kept private, so as not to influence the outcome of the council, but it was apparent that John, about to turn eighty-one, would not have long to live. The council would be his first, and last, grand project.

Thursday, October 11, 1962, dawned gray and drizzly, but the sun would soon arrive to reveal a spectacle quite literally medieval in its splendor. At 8:30, the bronze doors of the papal palace on the side of Saint Peter's Square swung open, and row upon row of bishops wearing white copes and miters made their way through the square toward Saint Peter's Basilica. Following this mass of white came a moving square of red: the College of Cardinals. And, finally, carried high in his papal sedan, came Pope John. He had never much liked riding in the chair, and to some observers he appeared uncharacteristically stiff and uncomfortable. As he was borne through the massed crowd of thousands, however, he began to wave and smile, then joyfully wept, as he blessed those in the square.

Once inside the basilica, under the bright glare of television lights, the bishops, cardinals, *periti,* and observers filled their tiered seats. Then, John—who had since departed from his sedan—walked the length of the broad aisle of Saint Peter's and took his place before the high altar on his throne. Pointedly, he had refused the usual grand papal throne, with red damask canopy, and ordered a less pretentious one. Mass was then said, in Greek and Latin, and the Sistine Choir sang hymns. After the Mass, the pope read, in accordance with canon law, a profession of faith dating from 1564: "I confess and hold the Catholic faith, outside of which no one can be saved."

The very pomp of the ceremony combined with such an un-ecumenical oath smacked of the old, ornate, and rigid Catholic Church, but this was far from the case with the pope's thirty-seven-minute address, delivered in Latin, which was one of the most extraordinary public statements ever made by a pontiff.

Peering through clear, gold-rimmed spectacles as he read in a "clear and resonant" voice that belied his age, John told those in the basilica (and, in effect, the world) that he was tired of hearing the words of naysayers within his own Holy Offices:

> In the daily exercise of our pastoral office, we sometimes
> have to listen, much to our regret, to voices of persons, who
> though burning with zeal, are not endowed with much
> sense of discretion of measure. In these modern times, they
> can see nothing but prevarication and ruin. They say that
> our era, in comparison with past eras, is getting worse. And
> they behave as though they had learned nothing from his-
> tory, which is nonetheless the great teacher of life. . . .
>
> We feel that we must disagree with those prophets of
> doom, who are always forecasting disaster as though the end
> of the world were at hand.

At this point in the pope's speech, bishops stole glances at Curia conservatives like Cardinal Ottaviani, Cardinal Giuseppe Siri, and numerous other cardinals, all of whom were seated close to John. It was rather extraordinary, like an American president scolding his cabinet at a State of the Union address. John did acknowledge the conservative cardinals among him, saying that the Church "must never depart from the sacred patrimony of truth received from the Fathers." However, he also pointed out that Catholicism had to "ever look to the present, to new conditions and new forms of life, introduced into the modern world, which have opened up new avenues to the Catholic apostolate."

John went on to say that he hadn't called the council to "discuss one article or another of the fundamental doctrine of the Church" but to take "a step forward toward doctrinal penetration." The repository of belief that all Catholics shared, the pope said, "should be studied and expounded according to the methods of research and literary forms of modern thought." This meant that historical and scientific studies of the Church and the Bible, in particular, were welcome in John's world; he wasn't afraid to shed rational light on the mysteries of faith.

He strongly emphasized two points in closing. "Nowadays," he said, "the bride of Christ prefers to make use of the medicine of mercy rather than severity. She considers that she meets the needs of the present day by demonstrating the validity of her teaching rather than by condemnation." There would be no thundering anathemas in John's council, in other words. And, finally, he made a call for unity: "The entire Christian family has not yet fully attained the visible unity of truth." The key to "the brotherly unity of all" would be love and charity.

This was an extraordinary message, one that let the world know where John stood—and where he wanted his council to go. *Aggiornamento* was at the heart of the speech, but the updating John had in mind was not merely window dressing or a useful slogan but a profound change in the way Catholics practiced their religion and viewed themselves and their Church.

John would not enter the council hall again until the second-to-last day of the session, in December, although he would watch the proceedings on closed-circuit television. But at the very beginning of the council, John made a point of making himself as visible as possible. On the evening of October 11, half a million people gathered in Saint Peter's Square. Catholic Action youths formed a huge cross around the obelisk at the center of the square, chanting, hoping the pope would come out on his balcony. He did, and was extraordinarily moved by the sight. Pointing up at the moon that shone down over the massive basilica, he told the multitude, "My voice is an isolated one, but it echoes the voice of the whole world. Here, in effect, the world is represented." Before he went back inside, he told the crowd, "Now go back home and give your little children a kiss—tell them it is from Pope John."

The next day, John met with seventy-nine diplomats, also invited to the council, to discuss international brotherhood and his desire for world peace, ironically a few months before the Cuban missile crisis. He also met with some of the 1,200 journalists covering the council, most of whom would leave at the end of the first session. Though he told them that "there are no political machinations here," few believed him.

Journalists in general grew frustrated with the secrecy surrounding council deliberations. Unless they were Italian reporters who were veterans of the Vatican beat, journalists could not figure out exactly what was happening, especially from the deliberately vague press releases sent out by the Curia-controlled Vatican Press Office, many of which were written, as one London periodical later complained, in "English so peculiarly outrageous that one hardly knows whether to laugh or cry."

John had made his humanist sentiments known to the world,

but the council couldn't avoid the classic battle between liberals and conservatives (though that is a great simplification). It is surely accurate to say some council fathers wished to preserve the status quo while others advocated change. Some have called the change agents *aggiornamentos*. Their influence was felt even before the council opened. Swiss theologian Hans Küng was already popular in progressive circles. Published in 1960, his book, *Council, Reform, and Reunion,* had been translated into eight languages. The American Jesuit John Courtney Murray led the charge for religious liberty, yet noted that "*the* issue under all issues" was the development of doctrine" (Murray's emphasis).

One progressive theologian, Father Jean Danielou, initially excluded by Alfredo Ottaviani from the council's preparatory commission, was later added to the council roster as a *peritus*— and a hugely influential one at that. John agreed to Danielou's exclusion from the preparatory commission but insisted on him for the council. Ottaviani also reluctantly accepted the liberals Yves Congar and Henri de Lubac. Other so-called progressives included Johannes Willebrands, a future cardinal, member of the Secretariat of Christian Unity and a leader in ecumenism and admirer of John Henry Newman; Karl Rahner; Gerard Phillips of Belgium; and the Belgian Edward Schillebeeckx.

Among the cardinals considered progressives were Augustin Bea; Achille Lienart, bishop of Lille; Julius Dopfner, archbishop of Munich and Freising; Josef Frings, archbishop of Cologne; and Franz Koenig, who had taught theology at Vienna and Salzburg and who brought Karl Rahner (whose writing had already been suppressed) as an adviser. Also, Bernhard Alfrink, archbishop of Utrecht; Paul-Emile Léger, archbishop of Montreal; Leon-Joseph Suenens, archbishop of Malines-Brussels—"northerners" in opposition to the southern European bloc.

Patriarch Maximos IV, who led the Melkite group, was a colorful sort of renegade. (One of his allies was Elias Zoghby, Melkite
partriarchal vicar for Egypt.) Maximos spoke French, not Latin,
did not wear the Western bishop's miter, spoke to Eastern patriarchs first, before he spoke to his Western counterparts, and as
early as May 23, 1959, suggested John establish the position of Secretariat for Christian Unity.

The conservative bloc included Cardinals Ottaviani; Ernesto
Ruffini; Giuseppe Siri; Irishman Michael Browne, former master
general of the Dominicans; Archbishop Dino Staffa, secretary of
the Congregations for Seminaries and Universities; Arcadio Lar-
rona, a Spaniard who was dubbed the soul of the opposition; Cardinal Rufino Santos of Manila; and Marcel-Francois Lefebvre,
archbishop (not cardinal) of Dakar and recently elected head of
the Holy Ghost Fathers. (Lefebvre, the most conservative of the
lot, would later lead a major schismatic movement in the 1970s
and beyond.) These members of the minority formed the international group of fathers, which included Geraldo de Proenca
Sigaud of Brazil and Luigi Carli of Segni, Italy.

The sign that the council would not be a rubber stamp for the
Curia, the pope, or anyone else came on Saturday, October 13,
1962, the council's first working-day session. The council was
broken down into ten separate commissions. Each commission,
formed in advance of the council by the ten preparatory commissions, considered a specific issue—such as potential changes in
the liturgy—and would then present draft proposals for decrees
(known as schemata), which the bishops at large would later deliberate and, eventually, make final. The council's first order of
business was to elect sixteen bishops to run each commission, a
total of 160 in all, although John appointed an additional eight
members to each commission. The Curia approved in advance of

the council the candidates for these highly important positions, expecting the assembled prelates to approve their choices.

But Cardinal Achille Lienart, bishop of Lille, France, offered a motion to delay voting for three days to learn more about the candidates. His motion was met with applause. This initial maneuver cannot be overestimated.

Much to the dismay of Cardinals Ottaviani and Tisserant, who presided over the day-to-day workings of the council, the bishops refused to vote for their approved list of candidates. Tisserant was forced to adjourn the meeting after only fifteen minutes, and when the session resumed on October 16, the bishops returned with their own list of candidates. "We found we were a council," an American bishop present said, "called here not as schoolboys, but rather to give a considered opinion."

The newspapers called it "The Revolt of the Bishops." With great glee, parodying the response of hand-wringing Italian cardinals like Ottaviani, one English archbishop wrote in his diary, "Scandalous! What a sight before the whole world!"

Both sides spent the next week jockeying for position. It soon became clear that the Council of Presidents, the body empowered to structure the sessions and provide the agenda, was in reality ineffectual and even detrimental. For instance, Cardinal Tisserant and the Council of Presidents had scheduled the liturgy as the first topic up for debate—a subject that would be sure to seriously divide the conservatives and liberals—rather than something less controversial that might allow them a chance to see that they could cooperate.

Cardinal Giovanni Montini, the future Pope Paul VI, wrote a letter to John in which he complained, "The choice of the liturgy as the first topic for discussion, although it was not placed first in the volumes distributed to us and although there was no need for it to come first, confirms the fear that there is no pre-established plan." Montini offered to the pope a structure for the council that would address all-important issues in at least three two-month sessions, spread out over a three-year period. This was bad news both for John, whose poor health naturally made him hope for a brief council, and for Cardinal Ottaviani, who also wanted a short council, with a Curia-approved agenda quickly decided upon, but Montini was prescient (it actually took four sessions and four years).

While the council worked, John went about his daily business, making appointments and visiting various parishes within the diocese of Rome. Word of his illness eventually leaked out; much of Rome gossiped that he was to have an operation on December 10, which was not true. On October 23—the day discussion of the liturgy began—John was involved in a geopolitical drama of great importance. Unhappy with the presence of American nuclear warheads in Turkey, Soviet Union Premier Nikita Khrushchev had secretly installed Soviet missiles within Cuba; these missiles were discovered by American reconnaissance flights. President John F. Kennedy announced a blockade on all Soviet ships heading for Cuba, and the world waited tensely for a nuclear war to begin.

As one of a number of backdoor options open to him, Kennedy reached out to *Saturday Review* editor Norman Cousins, who was known to have connections with both the Russians and the Vatican. Kennedy asked him if he would be willing to make contact with the Vatican, and Cousins, through an intermediary, was able to get a message through to Pope John, asking him to do anything in his power to help persuade Khrushchev to stand down from the brink. John did two things. In his weekly radio address on October 24, he urged both countries to show restraint: "The pope always speaks well of those statesmen, on whatever side, who strive to come together to avoid war and bring peace to humanity." Later, in a private note to the Soviet embassy in Rome, he delivered a more pointed message: "To promote, encourage, and accept negotiations, always and on every level, is a rule of wisdom that draws down heavenly and earthly blessings."

Khrushchev was already seeking a way out of the situation; the pope's message allowed him to present himself to the world as a peacemaker.

Ironically, however, war was also on the verge of breaking out at the Vatican Council as deliberations on potential changes in the liturgy began. It kicked off with an examination of the eight-chapter-long schema written by more progressive bishops and theologians. It argued that changing Church liturgy to keep abreast of modern times would bring more people into the fold, not fewer, as the conservatives, worried about losing the purity of traditional approaches, argued.

Jesuit Father John W. O'Malley, in his authoritative study, *What Happened at Vatican II,* puts the issue succinctly:

> The first issue protractedly debated, however, was the place of Latin in the liturgy, which intermittently occupied the council for several weeks. The issue, important in its own right, also had deeper ramifications. It was a first, awkward wrestling with the question of the larger direction that the council should take—confirm the status quo or move notably beyond it. The council resolved the question of Latin by taking a moderate, somewhat ambiguous, position. After the council that position got trumped by the most basic principle Vatican II adopted on the liturgy— encouragement of the full participation of the whole assembly in the liturgical action. This is a good illustration of a wider phenomenon of the council. Sometimes the inner

logic or dynamism of a document carried it beyond its original delimitations.

Some cardinals—Montini and Josef Frings of Germany, for instance—argued that changes in the liturgy embodied exactly what the pope had wished for in his council, an updating while keeping the sacred message of the Church the same. Two other cardinals indicated their frustration that the document as originally prepared had been altered before it was transmitted to the council session. Inserted by unknown hands had been a warning that this schema only dealt generally and theoretically with changing the liturgy and did not address the issue of local language versus Latin—that the practical application was ultimately up to the Holy See. Another section dealing with how the liturgy was in some ways biblically inspired was excised.

The bishops insisted that the original text be restored. They then indicated that what the majority desired was national and regional conferences of bishops who propose what liturgy changes might work in their own countries. Naturally, the pope would have to approve these suggested changes, as a way to safeguard the Church against false practices. The bishops knew—as did the cardinals of the Curia—that this might result in a certain decentralization of power within the Church hierarchy, since currently all changes to the liturgy had to go through the Sacred Congregation of Rites. This made the debate all the more fraught with tension.

Much of the discussion naturally focused on the use of Latin in the Mass. Cardinal Francis Spellman of New York indicated that he thought Latin should be continued, since it created a consistency and unity throughout the worldwide Church. Ironically enough, however, his own Latin was so poor that most of those present could not understand him, and he was forced to use a Latin translator to speak for him. In fact, many of those attending had problems understanding Latin well enough to follow the proceedings, and many eventually sought access to the translation services that Cardinal Bea provided for non-Catholic observers.

Influential Melkite leader Maximos IV Saigh, patriarch of Antioch and of all the Orient of Alexandria and Jerusalem, said, "All languages are liturgical, as the Psalmist says, 'Praise the Lord, all ye people.' . . . The Latin language is dead. But the Church is living, and its language . . . must also be living because it is intended for us human beings, not for angels." The debate, which lasted three weeks, from October 22 to November 13, encompassed fifteen sessions, 328 interventions from the floor, and 297 written interventions from the council fathers.

Cardinal James McIntyre of Los Angeles and other conservative prelates backed Cardinal Spellman. Archbishop Enrico Dante, secretary of the Congregation of Rites, argued forcefully that the Mass should continue to be said in Latin and complained that there was no mention in the schema about the veneration of relics. This brought an angry response from Bishop García Marquez of Columbia, who wondered aloud how much longer the Church would embarrass itself by worshiping relics like the Blessed Virgin's milk and Saint Joseph's sandals. These things, he said, "should be reverently buried and heard of no more." This had its humorous side—the cardinal presiding over the discussion had to cut him off for going on too long—but it showed the difference between the curial feeling in Rome and that of bishops out in the world.

Other prelates weighed in on the side of a vernacular Mass. One of the most influential was Cardinal Maurice Feltin of Paris, who depicted a scene in which a non-Catholic might find himself at a Mass unable to comprehend what was transpiring and thinking that this arcane and mysterious-seeming Latin rite was only for a certain in-group of people. How could this possibly help spread the word to those the Church had not yet reached? Other prelates pointed out that neither Christ nor his followers spoke Latin—indeed, the faithful did not even use Latin for the first 200 years of the Church's existence. And, of course, Latin was not used in Eastern Catholic churches.

This issue divided the bishops throughout Vatican II. It came to symbolize the vast difference in outlook between the curial prelates and the pastoral ones. Archbishop Pio Parente, secretary of the Holy Office, made an impassioned speech in which he told the assembly, "We are true martyrs at the Holy Office. We know how much patience, how much work, how much prudence is needed to prepare monita (warnings), decrees, etc. It's very hard work. You have no idea. And all this work is done in Latin, and a good thing, too."

Generally speaking, it is never a good idea to complain to others about how hard you are working (compared to them), and this little outburst did not go over well with bishops who labored in pastoral work year after year. The session had by this time broadened its liturgical subject to include the concelebration of the Mass—two or more priests celebrating at the same time, which emphasized the communality of the rite. Another important issue under discussion was the reception of the Eucharist by the faithful in the form of wine as well as bread (at that time, only the priest drank from the chalice).

The majority of bishops present spoke in favor of these changes—although some wanted the Precious Blood to be offered to communicants only on special occasions—as well as in favor of saying at least portions of the Mass in the vernacular.

Sensing a growing consensus, Cardinal Ottaviani rose on October 30 to head it off. "Are the Fathers planning a revolution?" he asked before coming down firmly on the conservative side of every issue. He said too many changes would confuse the laity. He claimed that concelebration of the Mass would turn it into a kind of theatrical production. Rather insultingly, he said that priests would not support this "new-fangled concelebration" once they learned that they would miss out on the Mass stipends often paid to priests to say Mass for a certain individual or intention.

The rule was that each speaker could only talk for fifteen minutes, but, carried away, Ottaviani continued to excoriate the liberals in the session, at which point Cardinal Bernard Alfrink of the Netherlands, president of that day's proceedings, interrupted him. "Excuse me, Eminence," he said, "but you have already spoken for more than fifteen minutes." The fact that a Dutch cardinal cut off the mighty Ottaviani in mid-speech was at once astonishing and heartening to the liberals in the session, who broke out into applause. Muttering angrily, *"Iam finivi, iam finivi!"* (I've already finished), the furious Ottaviani left in a huff. He would not return for two weeks.

PART III

Father of the Council

CHAPTER TWELVE

A New Pentecost?
(November–December 1962)

O n November 6, the Holy Father stepped in to the proceedings to change the rules allowing the discussion to be cut off by the president, effectively enforcing the ten-minute speaking limit. A week later, in another instance of his guidance, he inserted the name of Saint Joseph into the liturgy, to the protestation of the Orthodox and Protestants in attendance, who felt Joseph had already been sufficiently honored.

Although John refrained from attending the council meetings in Saint Peter's Basilica, the bishops and cardinals who were engaged in their historic debate felt his presence strongly. He met almost daily with groups of bishops from different countries, having what one prelate called "intimate and familiar discussions." He did not downplay the battles raging within the council sessions: "Yes, there's an argument going on," he told a group of French bishops. "That's all right. It must happen. But it should be done in a brotherly spirit. It will all work out. *Moi, je suis optimiste.*" (I, for one, am optimistic.)

Aside from these meetings, the pope spread the word about his true feelings through his close supporters. Cardinal Montini was one; he argued on behalf of John in one session that liturgical changes were necessary—and within Church tradition—because they would more effectively serve the needs of the laity. "It is," he said, quoting Saint Augustine, "better that we should be blamed by literary critics than that we should not be understood by the people."

It was in fact a clever tactic on the part of the progressive faction within the council to use the words and actions of revered Church Fathers against the traditionalists. One of the things that had so enraged Ottaviani was that the reformers often brought up the example of Pius XII when justifying their desire to make changes in the liturgy. (Pius had allowed Saturday evening masses to fulfill the Sunday mass obligation. He also changed the strict fasting for receiving Holy Communion.) "If you want to use the authority of Pius XII, use it not just when it agrees with you," he scolded liberal factions, pointing out that Pius was firmly against changing the Latin Mass.

But the progressives persisted in stealing the conservatives' own sacred cows and milking them for all they were worth. Cardinal Ernesto Ruffini, a staunch conservative and ally of Ottaviani, told the council that communion should not be offered under both species because the Council of Trent in 1551 had forbidden it. (Grasping at straws, he also claimed that it was unhygienic as well as inconvenient for busy priests.) But Cardinal Bea, John's inspired pick for the Secretariat for Promoting Christian Unity, turned the tables on Ruffini through an adroit use of historical research. True, the bishops at Trent had voted against communion with both bread and wine, but the vote had been close, eighty-seven to seventy-nine, which, he pointed out, would not meet the current council standards, in which a two-thirds majority was needed for approval.

And, he added, Pius IV's own representatives at Trent had approved allowing laypeople to receive the cup. In fact, two years after Trent, Pius IV himself had allowed the change in certain German dioceses.

Dogma, Bea and the others maintained, was immutable, but its expression was not.

On Sunday, November 4, the feast of his beloved Saint Charles Borromeo and the fourth anniversary of his coronation, John attended Mass at Saint Peter's. Cardinal Montini celebrated, but John gave the homily. The pope spoke fondly of Charles Borromeo, praising his contributions "to the renewal of Church life through the celebration of provincial councils and diocesan synods." After beginning his homily in Latin, he delivered the majority of it in Italian, a language, he said, "better understood by the greater part of this assembly; better understood by the crowds of faithful who have come to celebrate the anniversary of their Pastor and Father."

He finished with a gentle reminder to the Church fathers present:

> It is perfectly natural that new times and new circumstances
> should suggest different forms and methods for transmitting
> externally the one and same doctrine, and of clothing it in
> a new dress. Yet in the living substance is always the purity
> of the evangelical and apostolic truth, in perfect conformity
> with the teaching of the holy Church, who often applies to
> herself the maxim: "Only one art, but a thousand forms."

That is, *ars una, species mille*. Although couched indirectly, the meaning of the pope's homily was not lost on the Vatican Curia, who saw that they were steadily losing ground. An air of confusion and excitement grew at the council; change was in the air, but no one knew, not quite yet, what was going to happen.

During some of the longer and more boring speeches at council sessions, the bishops would retreat to a little coffee shop, one of two that the pope had ordered set up in the external corridors leading from Saint Peter's side entrances. Knowing his men, John had said that if there was no place to smoke, "the bishops will be puffing under their miters." There they chatted with all types of council observers, non-Catholics included. With conversation at times far more scintillating than the speeches ringing out in council sessions, the prelates often had to be urgently summoned two or three times to return to work.

In the middle weeks of November, the council began to examine a number of issues relating to the reading of the Divine Office, the Church calendar (the liturgical year), sacred ornaments and vestments, Church music, and Church art. Some of the discussions became long-winded, forcing wholesale retreats to the coffee bar. One bishop made a tiresome speech on the Virgin Mary and was cut off by Cardinal Ruffini, presiding for the day, who said, "One does not preach to preachers."

Bishop Petar Cule of Yugoslavia did just that, however, in a nervous, stuttering voice, pleading for Saint Joseph to be included in the list of saints recited in the canon of the Mass, at which point an impatient Ruffini interrupted, "Complete your holy and eloquent speech. We all love Saint Joseph and we hope there are many saints in Yugoslavia." After watching this on closed-circuit television, John stepped in to defend Cule, since he knew something most of the other prelates did not—that Cule's halting manner came from being tortured and nearly killed by communists in Yugoslavia.

There were, however, lighter moments too. When Cardinal Spellman and Cardinal McIntyre, the two most prominent American Church leaders opposed to the use of English in the Mass, voted in favor of allowing priests to read their breviary (the Liturgy of the Hours, or book of daily prayers) in English, an Italian archbishop slapped his brow in mock horror. "These Americans," he said. "Now they want the priest to pray in English and the people to pray in Latin!"

Since the Council of Trent in the sixteenth century, the Catholic Church had placed more emphasis on tradition—on the body of religious truths as passed down by the apostles. The Bible and tradition had been seen as two separate, independent sources of revelations, with the scriptures being given lesser emphasis, not the least because of the Protestant emphasis on them.

In the years leading up to Vatican II, however, many theologians felt that the Bible and tradition must properly be viewed as whole (as indeed they were in the early Church)—two roads stemming from the same source and leading to the same place. As the influential theologian Father Yves Conger put it, "There is not a single dogma which the Church holds by Scripture alone, nor a single dogma which it holds by tradition alone." This was far closer to the Protestant view and one held by many Catholic scholars. The Theological Commission, however, ignored this topic when it wrote its presentation to the council.

The council reached another crisis on November 14 when it began to debate the next schema, *De fontibus* (On the Sources of Revelation), which had been prepared by the Theological Commission chaired by Cardinal Ottaviani. If the liturgy was an expression of Catholic belief, revelation—how God spoke the truth to humankind, through tradition and scripture—was the very essence of it. Protestant churches believed that God revealed himself through the Bible alone.

An American Protestant observer wrote, "The dam broke." During the opening discussion on *De fontibus,* there were a couple of interventions, with Maximos IV Saigh and Joseph De Smedt of the Secretariat of Christian Unity asking for a new type of discourse.

As the session began on November 14, Ottaviani—finally back after leaving angrily two weeks before—announced the schema (which those present had already read) and was followed by Monsignor Salvatore Garofalo, a biblical scholar who summarized its contents and expressed the opinion, held by Ottaviani and the conservative faction, that the Church needed to condemn error by its own theologians in order to keep its doctrine pure.

Then Cardinal Ruffini declared that, if nothing else, the schema should be approved by the council simply because there was no other document available to take its place: "It would be as though a calamitous storm suddenly swept away the foundations of a great building." This was absurd—there were numerous theological documents already circulating, unofficial ones to be sure, that promulgated the opposing view, so many that Ottaviani felt forced to inveigh against them as "unauthorized documents that were against the rules and only caused trouble."

There were a number of problems that the progressive faction had with the schema as written by the Theological Commission. One common one, as Cardinal Achille Liénart of France rose to announce, was that it was "a cold and scholarly formulation, while Revelation is a supreme gift of God—God speaking directly to us." The schema as presented dealt not with the "gift" of God, but with what most people would consider technicalities.

Too, it was dogmatic in tone, lacking a certain generosity of spirit, which ran counter to John's wish for a pastoral council as expressed in his opening remarks. Liénart was followed by Cardinal Frings, who attacked the document's lack of ecumenism: "What is said here about inspiration and inerrancy [of the scriptures] is at once offensive to our separated brothers and harmful to the proper liberty required in any scientific theory."

Despite the best efforts of conservatives like Ruffini and Siri, liberals continued to speak against the document. Cardinal Joseph Ritter of Saint Louis walked to the microphone and announced, *"Rejiciendum est!"* (It must be rejected.) He went on to attack the conservative draft in harsh terms: "What a tedious and unrealistic attitude it betrays toward the Word of God which we call the Scriptures."

Later, the eloquent Cardinal Bea capped all the progressive discussions with this cogent point: "What our times demand is a pastoral approach, demonstrating the love and kindness that flow from religion." The schema, he continued, "represents the work of a theological school, not what the better theologians think."

Ottaviani and Ruffini were so upset at the objections raised by the liberal faction that they complained directly to the pope. John, however, was unmoved, perhaps even a little amused at the debate. He went on to tell them that things had been worse at the Council of Trent, where an enraged Latin bishop tore at the beard of a Greek prelate. Their protests were for naught, however. On November 20, the bishops voted to strike down the schema, 1,368 to 822, just short of the two-thirds majority required for motion. Both liberals and conservatives were discouraged. Not until the third session of the council would the debate be resolved. Scripture, tradition, and the *magesterium* (the Church's teaching authority) would be defined and reaffirmed as the sources of divine revelation for all—and for all time—in *Dei verbum,* the Dogmatic Constitution on Divine Revelation.

The first session of the council was due to end December 8, and what had they accomplished but a good deal of bickering? In terms of the Theological Commission's schema, they were now required to go back to Section One and begin debating all over again. How were they to break this logjam?

Later that night, Cardinal Paul-Emile Léger of Montreal, a former close friend of John from his days as Paris nuncio, visited the pope. The fighting going on in session disturbed Léger, and he told the pope, as he later recalled, that "piercing thorns were tormenting my soul." John told him that he should "go forward. Do what your heart tells you."

The influential Léger returned to the council for the discussion of the upcoming schema on the Church's relationship to mass media, movies, and television and forcefully made his feelings known.

In the meantime, unbeknownst to members of the council, John—fully aware that if discussion of Church issues stalled at this stage, his council would most likely stall as well—set up a special committee to deal with the failed schema on Revelation. In a moment of inspiration, he paired the conservative Ottaviani and the liberal Bea to co-chair the committee, making a strong point, in his role as Holy Father, that it was time for the squabbling prelates to get along. He directed Ottaviani and Bea to rewrite the schema, giving the liberal faction a symbolic victory. As John W. O'Malley, S.J., explains in *What Happened at Vatican II,* the committee "dealt a heavy blow to the control the Doctrinal Commission was trying to exert" on the Vatican Council.

On November 21, as bishops assembled to vote to suspend debate on *De fontibus,* word spread about the new committee, to the relief of those in favor of and opposed to the schema alike.

This would be the only time during the Second Vatican Council that John would enter directly into the proceedings, but his intervention arrived at a critical time. By stepping in when he did, John changed the course of the entire Second Vatican Council. The importance of his action can hardly be exaggerated.

At the very end of November, the official Vatican paper, *L'Osservatore Romano,* admitted for the first time what almost everyone had divined—that Pope John was ill. The paper wrote that he had been forced to cancel papal audiences because "the symptoms of gastric disturbance were getting worse; for some time, the Holy Father has been on a diet and undergoing medical treatment that have led to rather severe anemia." Nowhere was the word *cancer* mentioned. In fact, the pope's doctor had been giving him cobalt-ray treatment for his stomach cancer, which in turn had caused the anemia. Treatments had failed; John now understood that he had less than a year to live, according to his doctor's estimate.

The Holy Father was struggling—he had to be helped to appear at his window for Angelus on December 2—yet he was determined to write the encyclical that would become *Pacem in terris,* which had come to him in a moment of inspiration during the Cuban missile crisis.

Despite his poor health, he continued to avidly follow the doings of the council as it reached its final days. The schema up for discussion as November ended was one on the Unity of the Church, which had been prepared by the Commission for the Oriental Churches. As the name of the commission implied, it was concerned with the ways in which union might be achieved with the Eastern Orthodox faith.

There were two other schemata on unity presented as well: one from the Theological Commission concerning itself with closing the gap between Catholics and Protestants, and one from Bea's Secretariat for Christian Unity, which was concerned with ecumenism generally. Many questioned why all three of these efforts did not stem from the Secretariat for Christian Unity, under whose purview they might naturally fall; the answer was that a rivalry existed between commissions, especially that of Ottaviani's Theological Commission, which considered itself first among equals.

But John, who was the true first among equals, had planted the seed of ecumenism, and it would not be dislodged or uprooted. It would bear fruit in the third session of the council, two years later.

It was, of course, ironic that there was such disunity apparent in a discussion of unity. Naturally, given the tenor of the council, Ottaviani's schemata were overwhelmingly defeated as being too authoritative and not ecumenical enough. Seeing that there was little he could salvage, Ottaviani introduced his next schema, ominously titled "On the Nature of the Church Militant." A seemingly upbeat Ottaviani told the assembled:

> I expect to hear the usual litanies from you all: it's not
> ecumenical and it's too scholastic, it's not pastoral and it's
> too negative, and similar charges. This time I will make a
> confession to you: those who are accustomed to say "Take
> it away and replace it" are already poised for battle. . . . All
> that remains for me is to fall silent for, as Scripture says,
> where no one is listening, there is no point in speaking.

Presented in a jocular tone of voice, these words caused laughter and applause, and Ottaviani left the microphone smiling, an unusual sight, given his fortunes during the council. However, the Theological Commission's schema reasserted the mystical body of Christ as associated only with the Catholic Church— all other Christians (and non-Christians) were out of luck. The schema portrayed the Church "militant" (essentially, the faithful of the Church) as a kind of pyramid, a hierarchy with laypeople at its base, followed by priests and pope. In expounding on "states of evangelical perfection," Ottaviani's schema went on to say that true holiness could only be achieved by those who had taken vows of poverty, chastity, and obedience, although the laity did have certain rights and responsibilities within the Church—to live according to gospel precepts, to preach and to teach, to raise Christian families, and to be "vicars of Christ" in the secular world—which added up to a kind of vocation over and above their secular calling.

Perhaps the best response to this, most observers agreed, came from Bishop Emile Jozef de Smedt of Brugge, Belgium, who deconstructed the schema into three areas of criticism. One was "triumphalism," by which he meant priests seemed to have a leg up on getting to heaven compared to the laity. The next was "clericalism"—that big pyramid with the pope and the clergy on top of the laity. Finally, there was "juridicism." Instead of dealing with the pastoral Church, de Smedt argued, the schema split hairs, inhabited by the petty spirit of legalism. This last, of course, was a criticism of almost every document that came from the Theological Commission. And, as with these other schemata, the one on the Church militant went nowhere.

So, it would seem, after a century or more of conservative ascendancy in the Church, the progressive faction was triumphant.

But was it? What, exactly, did this first session, now rapidly drawing to a close, succeed in doing? Had the notion of *aggiornamento* been satisfied? Was the Church updated?

Cardinal Giovanni Battista Montini put his finger on the worries felt by prelates on both sides of the aisle when he wrote, "A vast amount of excellent material has been brought together, but it is too disparate and uneven. . . . A central and controlling idea is needed to coordinate this immense material." Later, in a speech to the assembled Church fathers on December 5, Montini told his peers, "Some are afraid that the conciliar discussion will be endless and that instead of bringing people together it will divide them even more. But that will not happen. The first session has been a running-in period. The second will progress much more swiftly."

The second session was scheduled to begin in May 1963, but the bishops convinced John to delay it until September so they could spend more time in their home dioceses. Though fairly certain he wouldn't live long enough to open the next session, he remained hopeful. He told a crowd of onlookers that "good health, which has threatened for a moment to absent itself, is now returning, has actually returned."

On December 4, he blessed pilgrims to Rome from his window. A vast crowd gathered, not just pilgrims, but the bishops of the council, who played hooky, as it were, to see the pope. When he appeared in the window, he was greeted by a huge roar of approval. He waved a hand for silence, and then told the rapt audience, "My children, Divine Providence is with us. As you see, from one day to the next there is progress, not going down, but in coming up slowly. . . . Sickness, then convalescence. Now we are convalescing."

Waving an arm in a gesture that encompassed the whole of the square, John alluded to what he thought was so important about the council: "What a spectacle we see before us today—the Church grouped together here in full representation—behold its bishops; behold its priests; behold its Christian people. A whole family here present, the family of Christ!"

Angelo Roncalli had remained close to his large, extended family; Pope John was equally attached to the family of Christ. Despite all the infighting and the politics, it was what he wanted the Church fathers in the council to understand, and the reason his sympathies lay with the more progressive views of Cardinal Bea and others—they wanted to open up the Church, to welcome all with open arms, as Marianna Roncalli once welcomed beggars to her overcrowded table for a bit of polenta.

On December 7, the last plenary meeting of the first session was held, and the pope surprised everyone by slipping in through a side door and walking, unaided, to take his seat. He was greeted by applause, prayed with those present, and then quietly left. On December 8, the final day of the first session of the Second Vatican Council, and the Feast of Immaculate Conception, John addressed the assembly. He was pale but vigorous and seemed to get stronger as he pushed his glasses up on his nose and began to speak in Latin:

> Now that the bishops of the five continents are returning to their beloved dioceses . . . we should like to dwell a little on what has been done so far, and to map out the future. . . .
>
> The first session was like a slow and solemn introduction to the great work of the council. . . . It was necessary for brothers, gathered together from afar around a common hearth, to make each other's close acquaintance; it was necessary for them to look at each other squarely in order to understand each other's hearts. . . .
>
> [The second session] will be a new Pentecost indeed, which will cause the church to renew her interior riches and to extend her maternal care in every sphere of human activity. . . . In this light we look forward to your return, we salute all of you "with a holy kiss," while at the same time we call

down upon you the most abundant blessings of the Lord, of which the apostolic blessing is the pledge and the promise.

When he finished, John stepped down from the platform and walked out the side door of the basilica. The first session of the council—the only one he would see—was over. While the great burning issues of the Church—the liturgy, the reach of ecumenism, the true role of laity—had yet to be decided, John's huge accomplishment was to make the far-flung bishops realize that the Church was truly "catholic," truly universal. When he addressed the crowd on December 8, he told them, "Each man must feel in his heart the beat of his brother's heart." And it was not only the bishops who profited by this, but an entire world that had now begun to look at the Catholic Church in a very different light indeed.

The Second Vatican Council, in the long run, under the direction of Paul VI, did not tackle artificial birth control, clerical celibacy, and substantive reform of the Roman Curia (all of which Paul would address himself). But it did address matters such as the liturgy in the vernacular, the lay apostolate, the Church in the modern world, the role of clergy and religious orders, collegiality among bishops, the restoration of a permanent diaconate, Christian unity, and—of maximum consequence—the relation of the Catholic Church with other faiths, including Judaism.

After the bishops went home to their dioceses and the seats in Saint Peter's Basilica were taken down and the crowds disappeared from the piazza (relatively speaking, of course), John settled back into the business of being pope. Despite his poor health, he had a pressing issue to resolve. For seventeen years, the Soviets had held in dire imprisonment Metropolitan Josef Slipyi, the Orthodox archbishop of Lviv, Ukraine. Once again, Norman Cousins, the editor of the popular magazine *Saturday Review,* was able to effect a breakthrough; attending the council, he informed a few cardinals close to John that he had a meeting fixed with Nikita Khrushchev in mid-December, at which he would press for Slipyi's release.

John passed a message through Cousins to the Russian premier that he would view releasing the metropolitan as a sign of extraordinary good will. Cousins met with Khrushchev and then passed on a lengthy report of the encounter to John. In it, the *Saturday Review* editor said that the Soviet leader didn't know where Slipyi was at present. However, Khrushchev told Cousins, "I will have the case examined and if there are assurances that it will not be turned into a political case, I will not rule out liberation. I've had other enemies, and one more at large doesn't bother me."

Khrushchev was still thankful for the pope's intervention during the Cuban missile crisis, and even compared the two of them: "We both come from humble origins and worked the land in our youth." For Slipyi to be released, however, the Vatican would have to provide something to Khrushchev in return—a private back channel of communication opened up, which would serve the Russians well in case of another global emergency.

On December 19, Cousins had a forty-minute audience with John and delivered a message from Khrushchev wishing him good health. "I get many messages from people who are praying that my illness may be without pain," he replied. "But pain is not my enemy. I have memories, so many marvelous memories."

One of these memories was of being apostolic delegate to Bulgaria, of riding through the rugged backwoods to meet the ordinary people of the country. "The Russian people are a wonderful people," he told Cousins. "We must not condemn them because we don't like their political system. They have a deep spiritual inheritance which they have not lost. We can talk to them. We must always try to speak to the goodness that is in people."

In this spirit, on December 22, John personally typed a reply to Khrushchev that began, "Cordial thanks for the courteous message of good wishes. We return them from the heart in words that come from on high: Peace to men of good will." He also enclosed a copy of his Christmas address, which he delivered later that day, as well as a picture of the Virgin Mary and a copy of a prayer that John had recited ever since his seminary days.

Pope John spent the last days of 1962 working on *Pacem in terris* and resting as much as possible. *TIME* magazine had named him its "Man of the Year," and fulsome compliments came in from around the world, both for the pope himself and for the spirit of unity fostered by his great council. Still hoping for Slipyi's release, John wrote in his diary:

December 26. A calm Saint Stephen's day. The liturgy
made a great impression on me. My spirit continues to be
concerned with whatever it is that the Lord is mysteriously
doing. Is not this Kroucheff—or Nikita Khrushchev as he
signs himself—preparing some surprise for us? After a long
meditation last night . . . I got out of bed and then, kneeling
before the crucified Lord, I consecrated my life and the final
sacrifice of my whole being for my part in this great under-
taking, the conversion of Russia to the Catholic Church. At
noon during the general audience in the Sala Clementina,
still under the same inspiration, I put great fervor of heart
and lips into the prayer, *Domine, tu scis quia amo te.* (Lord,
you know that I love you; John 21:17.)

CHAPTER THIRTEEN

Peace on Earth
(January–April 1963)

The initial session was over and the bishops were back in their dioceses, but on the very first day of the new year of 1963, Pope John XXIII was hard at work reminding them that their labors had only just begun. Awakening at 4:00 A.M. on January 1, John began work on a letter addressed to the "Bishops of the Council." In-between sessions, he wrote, should be "the apple of your eye," and even the most "urgent pastoral work" should not supersede it. The bishops must respond quickly to requests to review drafts of council documents and, in general, prepare themselves to hit the ground running in September.

Published on January 6, on the Feast of the Epiphany, the letter was a pep talk of sorts. John extolled the work done in the first session; even with all its flaws, the council had brought light and air on numerous subjects that needed addressing within the Church. Their early work, he told them, had let the world see that the Church was open to change and brotherhood. The

Coordinating Commission he had created to help sort out the profusion of draft texts would meet in a few weeks; John let the bishops know he considered it to be the guiding light of the next council session.

He also told the prelates that they should consider themselves prime movers and decision makers. Although the pope needed to approve everything, "it is up to the bishops to supervise, according to the rules, the free development of the council."

Led by Cardinal Leo Joseph Suenens of Belgium, the Coordinating Commission met from January 21 to 27. Suenens, a progressive who fully supported John's program, simplified matters by reducing the number of proposed schemata from seventy to twenty and dividing them into internal and external Church issues.

The next council session would tackle universal matters like marriage and the family, social justice, and the community of nations, among other issues. Suenens also wanted more secular experts involved in preparing documents for the September council session. John approved Suenen's proposal, happy that a crucial outline was in place.

For the last six months of his life, Pope John was concerned with making sure that matters ran smoothly after his death. He also wanted to set the record straight. The day after the last meeting of the Coordinating Commission, he wrote a formal letter to Monsignor Loris Capovilla, his faithful aide—the man who perhaps knew John better than anyone else.

As was often the case, John awoke early. "Dear Monsignor," the letter begins:

At four this morning I was awake and looking over conciliar material when it struck me that it would be good to think of a future "historian" of the great event that is under way, and that he will have to be chosen with care.

I think that the obvious witness and faithful exponent of "Vatican II" is really you, dear monsignor: and in so far as a *mandate* can come from me—pope of the council, alive or dead—you should be authorized to accept that task as the Lord's will, and do honor to it, which would also be an honor for holy Church, a pledge of blessings, and a special reward for you on earth and in heaven.

John wanted to prepare for a world after his death, when Loris Capovilla might in fact need the imprimatur of the pope to proceed with his labors in the face of those who opposed the pope, and he also wanted someone to record his side of the story when it came to issues beyond the council, for which he foresaw criticism.

One of these issues was John's continued attempt to free Metropolitan Slipyi from his long captivity behind the Iron Curtain. He had just sent Monsignor Jan Willebrands to Moscow to negotiate Slipyi's release, knowing full well that conservatives both inside and outside of the Church decried his conversations with the Soviet Union. Among Italian politicians especially, the word was that John was soft on communism.

The pope realized that there were those in the Curia allied with political forces who actively wanted to see him fail. He told a sympathetic editor of a Catholic journal that he needed, in his final days, "to be extremely careful in everything I do to prevent the conclave after my death being a conclave 'against me,' because then it might make a choice that would destroy everything I have started out to achieve."

In February 1963, Metropolitan Slipyi, who had been in the Soviet gulag for eighteen years, was suddenly transferred to the

Hotel Moscow, where, to his astonishment, Monsignor Willebrands told him the extraordinary news that he was a free man.
The tall, bearded, seventy-one-year-old Slipyi, arrested by the
NKVD (the People's Commissariat for Internal Affairs, a forerunner of the KGB) in 1944 on trumped-up charges of collaborating with the Nazi puppet regime in the Ukraine and the real-life
inspiration for Morris West's bestselling novel *The Shoes of the
Fishermen,* was grateful—until he learned he was not allowed to
go to the Ukraine.

He balked, until Willebrands told him that it was the pope's
personal desire that Slipyi leave prison and come to Italy, even if it
meant permanent exile.

Their journey back was like something out of a Cold War spy
movie. They took trains through Vienna and Venice, but got off
50 miles north of Rome in order to duck the press. Both John and
Nikita Khrushchev wanted to avoid headlines—as Khrushchev
told Norman Cousins, the kind that read, "Bishop Tells of Red
Torture"—because it would damage both leaders with conservative factions within their bureaucracies. Loris Capovilla met Slipyi
in a car, took him to a nearby abbey, and then returned to Rome.
It was late, but Capovilla slipped an excited note under the pope's
door:

> Holy Father! I got back at midnight. Metropolitan Slipyi
> arrived safely. He is very grateful to your Holiness. He
> admired your gifts. He said: "If Pope John in his goodness
> hadn't brought this off, I wouldn't have lived much longer."

The next day, February 10, the news appeared in *L'Osservatore
Romano,* but Slipyi's location was kept a secret, and John announced it during the dedication of a new seminary: "From East-

ern Europe there came last night a moving and consoling gift for which I humbly thank the Lord." In a private audience that evening, the metropolitan knelt before John and thanked him for his release. "You have pulled me out of the well," he told the pontiff.

They prayed together and Slipyi, who had written courageously during his captivity in essays that had been circulated in samizdat format, gave John a map of all the prison camps in the gulag, a map that John had with him at his death, and on whose margin he wrote, "The heart is closer to those who are further away; prayer hastens to seek out those who have the greatest need to feel and be understood."

Slipyi's release was a part of the Soviet "thaw" of the late 1950s and early 1960s, an easing of tensions that Kremlin hard-liners did not appreciate. For Khrushchev to do what he did, John said, constituted "the greatest political heroism," especially because, as the Soviet premier had feared, newspapers around the world did publish headlines decrying "Red torture." But it was John's overwhelming desire—as expressed in his interactions with the Soviets, in his preparations for the council session that would surely occur after his death, and in his encyclical *Pacem in terris,* still in development—to bring the world together rather than push it further apart.

John's note about how "the heart is closer to those who are further away" also applied to his own family, whose visits with him had been unavoidably cut short during his papacy. In a letter he wrote to his eldest brother, Zaverio (typing it himself), the pope addressed the family as a whole. Once again mentioning his death, he told the Roncalli clan:

Be of good heart! We are in good company. I always keep by my bedside the photograph that gathers all our dead to-

gether with the names inscribed on the marble: grandfather
Angelo, "Barba" Zaverio, our revered parents, our brother
Giovanni, our sisters Teresa, Ancilla, Maria, and Enrica.
Oh, what a fine chorus of souls to await us and pray for us! I
think of them constantly. To remember them in prayer gives
me courage and joy, in the confident hope of joining them
all again in the everlasting glory of heaven.

I bless you all, remembering with you all brides who have
come to rejoice with the Roncalli family, and those who have
left us to increase the happiness of new families, of different
names but similar ways of thinking. Oh, the children, the
children, what a wealth of children and what a blessing!

Almost everything John did during this time had a valedictory
air to it.

During the Lenten season he visited parishes throughout
Rome, where crowds gathered to see him. People wept and ap-
plauded him, realizing that this might be the last time they
might get a glimpse of the man they called "good Pope John."
In some quarters, elections were under way, but campaigns were
suspended when the pope came through—whatever political differences divided conservatives from liberals, they recognized in
John a genuinely holy man, as did his many visitors during this
time. He could be a little scattered as well. During an audience
with one Protestant minister, he forgot that he was not talking to
a priest and urged the man to pray to the Virgin Mary—but these
moments of vulnerability only made people love him more.

On March 1, the pope learned that he would be awarded the
Balzan Prize, an international award for people who have made
outstanding contributions to humanity. The citation read that the
pope was being honored for "his activity in favor of brotherhood

between all men and all peoples, and his appeals for good will in his recent diplomatic intervention," a reference to his role in the Cuban missile crisis.

The award was sniffed at by certain members of the Curia, who, according to one observer, "held it was undignified for a pope to be receiving a prize at all." Some also felt John was at the center of a growing "cult of personality." This latter criticism may have been true.

There had seldom, if ever, been a modern pope with a more personal touch than John XXIII. One day, he asked his longtime personal assistant Guido Gusso to bring Gusso's son to visit him. Like Capovilla, Gusso was one of the few people to interact with John on a daily, intimate basis. He was concerned about the pope. He later remembered that he cried out of sight of the pope as he witnessed the cancer's quickening progress through the pontiff's body.

"I lost all interest in food and drink," he later remembered. "All I could do was smoke cigarettes, four packages a day, so that now I cannot even bear the smell of them. In those seven months that the Holy Father was dying, I lost thirty-five pounds."

Recognizing Gusso's discomfort, the pontiff requested the visit. When Gusso brought in three-year-old Giovanni, the boy and the pope had difficulty understanding each other, since the child spoke in the Veneto dialect. However, he did tell the pope he wanted to become a priest, at which the affable pontiff laughed and replied, "You are too handsome. When the time comes, you will want to marry."

It was a moment Gusso would always treasure. The pope, in a great deal of personal pain himself, had reached out to those who loved him, knowing the memory would be a comfort to them.

Such personal moments, however, took a backseat to international controversies, such as the one concerning Nikita Khrushchev's son-in-law, Alexis Adzhubei, editor of *Izvestia,* the Soviet Union's official newspaper. Adzhubei married Khrushchev's daughter, Rada, in 1952. Since then, he had gained greatly in power and influence, becoming a member of the Russian premier's inner circle and a kind of traveling emissary. In fact, President John F. Kennedy granted Adzhubei the first interview an American president had ever granted to a Soviet journalist, a two-hour affair that took place with much fanfare at the Kennedy compound in Hyannis Port in Massachusetts.

On February 28, Adzhubei and Rada came to Rome with the express desire to visit the pope—the journalist said that he had a gift for John from his father-in-law. Since John's contacts with Russia remained controversial, he decided to ask Cardinal Ottaviani whether he should meet with Adzhubei. Ottaviani advised against it, arguing the Russians would use the visit as propaganda. But John said, "I would be breaking my word and condemning all my previous behavior if I refused to see someone who has courteously and sincerely asked to see me in order to bring a message and a gift."

But, finally, after about a week of wrangling within the Vatican, Adzhubei agreed to a meeting after John's general audience at the Vatican with selected journalists. Adzhubei and Rada joined John in the papal library, where he spoke to them in French (which Rada understood well) and then sat down on a couch with the couple on either side of him. The discussion continued with the aid of an interpreter, Father Alexander Koulic, who also kept notes on the visit. Adzhubei asked the pope what the pontiff thought about establishing diplomatic relations with Russia. The pope answered diplomatically, relating the story of Genesis: "The Bible says that God created the world and on the first day he created light. Then creation went on for another six days. But the days of the Bible, as you know, are whole epochs, and these epochs last a very long time."

The pope told Adzhubei that Russia and the Vatican were on the very first day, as it were, and that things looked promising for the creation of peace, but that it would not do to rush things: "We must go gently, gradually in these matters, preparing minds," he said. "At present, such a move would be inopportune." The meeting ended with the couple and the pope exchanging minor gifts. Adzhubei asked if he could publish an account of the meeting, and John replied no, which the Russian journalist accepted in good grace.

And that was it—an innocuous exchange. But it once again highlighted the situation John faced within the Holy Office. John had intended that Loris Capovilla should write an account of the meeting and publish it in *L'Osservatore Romano,* but Ottaviani and the Holy Office put their collective foot down. They would not hear of it. John was so upset at their flagrant act of disobedience that, on March 20, he dictated an extraordinary "note for history":

> The absolute clarity of my language, first in public and then
> in my private library, deserves to be known and not with-
> held on some pretext. It should be clearly said that the pope
> has no need to defend himself. . . . When it is known what
> I said, and what he [Adzhubei] said, I think people will
> bless the name of Pope John. Everything should be carefully
> noted down. I deplore and pity those who in these last few
> days have lent themselves to unspeakable maneuvers.

These were harsh words, but the pope was tired of Vatican careerists playing politics. (The immediate concern of Ottaviani and others was that, with Italian elections due to take place at the end of April, the pope would look like he was sympathetic to communism, which in turn might persuade Italian Catholics to vote for communist candidates.) But it was frustrating for John, who had once spoken sarcastically to a friend about the amount of "freedom and sovereignty" a pope truly had.

It's ironic that John would dictate this note just as he was getting ready to send his greatest encyclical letter, *Pacem in terris,* to press. Eleven days later, on March 31, he signed the first five copies in front of television cameras. It was due to be released on Holy Thursday, April 11, a day that John had specifically chosen. On that day, despite being in a good deal of pain, he spoke to the diplomatic corps:

> I'm glad that the encyclical has been published today, the
> day on which the lips of Christ pronounced the words,
> "Love one another." For what I wanted to do above all was
> to issue an appeal to love for the people of this time. Let us
> recognize the common origin that makes us brothers, and
> come together!

Pacem in terris (Peace on Earth)—John's eighth and final encyclical—is the most catholic of papal encyclicals, universal in its appeal. It is also the most "secular" and influential, with debate and impacts from it continuing to echo today. The 20,000-word document is bold, comprehensive, and ambitious. Its substance and mandate changed the political discourse of world leaders in its wake. Although popularly known for its three-word Latin title (as are most such documents), the full title reads, "On Establishing Universal Peace in Truth, Justice, Charity [Love] and Liberty." Its "peace on earth" message derives from the gospel tidings of the nativity. In a stunning departure from all other encyclicals, the pope addresses it not only to bishops and the Church faithful but also to "all men of good will."

John conceived the idea for *Pacem in terris* during the Cuban missile crisis, when he helped Khrushchev, through intermediaries, continue negotiations with the United States. To underscore these conciliatory moves, John sent an advance copy of *Pacem in terris* to Khrushchev in the Kremlin.

One can argue that the pope's unique background in diplomacy set the stage for such a far-reaching and ecumenical epistle. His previous postings in Bulgaria, Greece, Turkey, and France taught him the language of mutual respect and a vital appreciation and sensitivity toward diverse cultures, especially non-Catholic and even non-Christian and atheist.

Pacem in terris made quite a splash. Even before its debut, word of its contents sparked news stories. Nearly a week before it came out, a story in the *New York Times* previewed its theme of world peace. When it did come out, it garnered front-page and broadcast attention worldwide. Curiously, the *Times* seized on one component of the encyclical for page-one headlines, announcing that the pope hoped for a "world nation to guard peace." Its lead story proclaimed, "Pope John XXIII proposed in an encyclical today the establishment of a world political community or public authority, a kind of super-nation to which all countries should belong. Its aim would be to insure peace." The story added, "He made it clear that this new world organization should not be in contrast to or competition with the United Nations." The encyclical itself, however, paid only modest attention to this, in paragraphs 137 and 138.

Picture, if you will, the accompanying stories in the news at the time to understand the gravity of world tensions: A week earlier, in Geneva, the United States and Russia had agreed to a "hot line," a brand-new term for a communications link to reduce the threat of accidental war. At the same time, the North Atlantic Treaty Organization (NATO) powers were attempting to heal rifts over who would carry whose nuclear weapons in Europe. Berlin remained a flash point, with U.S.-Soviet talks at a stalemate. Rumbling in Southeast Asia merely hinted at the protracted and expanded war to come. And social upheaval in the United States was simmering when Dr. Martin Luther King Jr. was arrested at sit-ins in Birmingham, Alabama.

The encyclical itself comprised five main parts: order between men; relations between individuals and the public authorities; relations between states; relationship of men and of political communities with the world community; and pastoral exhortations. It logically and methodically built a case for world peace, beginning with order in the universe and order within human beings created in the image of a loving and peaceful God. As with all encyclicals, it buttressed its arguments with scriptural citations, references to Church scholars, and references to remarks of previous pontiffs, as well as John's own previous words.

Its introductory language hearkened back to the U.S. Declaration of Independence, the French Declaration of the Rights of Man from the Age of Enlightenment (so often criticized by Church authorities), and the Universal Declaration of Human Rights of the United Nations. Previous popes (even Leo XIII) had steered clear of such language. John, however, embraced the concepts of human rights and articulated such in understandable terms, not the usual coded, lofty language of pontifical pronouncements.

He noted that peace can never be attained "except by the diligent observance of the divinely established order."

Among the rights he enumerated were freedom of speech and access to information, the rights of families, the right to private property, and the rights of workers to earn a living wage (which did echo Leo XIII's *Rerum novarum*). But his discussion of rights was always balanced by a discussion of social obligations geared toward the common good.

He may be considered prescient in his view of women, highlighting their importance to the family and in the economic sphere. Foreshadowing feminist positions that would become more commonplace within the coming decade, he wrote:

The part that women are now playing in political life is everywhere evident. This is a development that is perhaps of swifter growth among Christian nations, but it is also happening extensively, if more slowly, among nations that are heirs to different traditions and imbued with a different culture. Women are gaining an increasing awareness of their natural dignity. Far from being content with a purely passive role or allowing themselves to be regarded as a kind of instrument, they are demanding both in domestic and in public life the rights and duties which belong to them as human persons.

Among his bolder propositions was his declaration that each state must have a public constitution to define the roles of political leaders in their relations with citizens. He called for the elimination of racial discrimination and the need for more advanced nations to make greater contributions toward developing nations. To drive home the importance of social justice among nations, John used a pithy epigram of Saint Augustine: "Take away justice, and what are kingdoms but mighty bands of robbers?"

Pacem in terris made a number of powerful personal appeals— for the protection of minority populations and refugees, for instance. But John saved his most impassioned pleas for nuclear disarmament. He belittled the squandering of resources on the arms race and even obliquely touched upon the unknown environmental impacts of nuclear testing. Recognizing the need for spiritual transformation at the heart of peace, he said that disarmament must "reach men's souls."

In one passage, John spoke prophetically of the emergence of the sort of global economy we witness today. "National economies," he wrote,

> are gradually becoming so interdependent that a kind of world economy is being born. . . . Each country's social progress, order, security and peace are necessarily linked with the social progress, order, security and peace of every other country. It is clear that no state can fittingly pursue its own interests in isolation from the rest, nor, under such circumstances, can it develop itself as it should. The prosperity and progress of any State is in part consequence, and in part cause, of the prosperity and progress of all other states.

In words that could apply to the Internet, he noted the "profound influence" of science and technology:

> This progress is a spur to men all over the world to extend their collaboration and association with one another in these days when material resources, travel from one country to another, and technical information have so vastly increased.

This has led to a phenomenal growth in relationships
between individuals, families and intermediate associations
belonging to the various nations, and between the public
authorities of the various political communities.

This was not the first time John acknowledged the power and
veracity of mass communication—and its utility in spreading the
gospel. More would come from the council.

The pope's heralded suggestion for a "supernation" stemmed
from his finding that sincere efforts toward peace were not
enough; radical systemic changes were needed. His proposal for
a new form of public authority was not a rebuff to the United
Nations—far from it, since he offered fulsome praise for that
organization—but a plea for nations to work toward the common
good to avert catastrophe, as they had during the Cuban missile
crisis.

In the closing pastoral exhortations of *Pacem,* John passionately
entreated his brothers and sisters to turn toward peace. No less
than five times he addressed people beyond the Church, citing the
longings in the hearts of "all men of good will." He underscored
the spiritual basis of peace in a passage reminiscent of a popular
Catholic hymn: "The world will never be the dwelling place of
peace, till peace has found a home in the heart of each and every
man, until every man preserves in himself the order ordained by
God to be preserved."

For the most part, world leaders received *Pacem in terris*
warmly. Breaking a precedent of silence toward papal encyclicals,
the U.S. State Department said, "No country could be more re-
sponsive than the U.S. to its profound appeal to, and reassertion
of, the dignity of the individual and man's right to peace, liberty
and the pursuit of happiness." In fact, one American diplomat in

Rome reportedly exclaimed, "It embodies everything the U.S. has been working for. We couldn't agree with it more."

On the other side of the political spectrum, the Soviet news agency TASS issued a brief summary of the encyclical, highlighting the pope's call for disarmament and attention to workers' rights, participation of women in public life, and equality of races and nationalities. Moscow's *Izvestia* newspaper made it clear the Kremlin viewed it favorably.

Communist leaders in Italy, Belgium, France, and Poland hailed it. A chorus of positive responses came from diverse quarters, including UN Secretary General U Thant and government leaders in France, Germany, England, and other countries.

In the United States, Protestant, Orthodox, and Jewish leaders were moved by the pope's broad appeal, as were civil rights groups. An official of the National Council of Churches termed *Pacem in terris* "a magnificent statement of world responsibility."

The head of the National Conference of Christians and Jews received it as a "masterpiece" and said it "will be widely understood as an emphatic rebuke to isolationists, narrow nationalists, racists and those who rely on mass retaliation." An executive of the National Association for the Advancement of Colored People called the encyclical "an unequivocal answer to those bigots and false prophets who like to justify racial segregation and other injustices by quoting the Bible out of context."

In theological circles, the prominent Jesuit theologian John Courtney Murray noted that *freedom* was added to the traditional papal pillars of truth, justice, and charity as the structural supports of peace. He saw *Pacem* as a shining example of John's *aggiornamento* and suggested the pope's endorsement of modern written constitutions was a papal first. Murray was careful to emphasize the pope's rejection of a Marxist view of history, which portrays human beings as determined by historical forces.

Pacem in terris still sparks admiration and commentary a half-century later. Was it naïve? Idealistic? Pioneering? The debate continues, falling along partisan lines. However, there is no doubt it was a turning point in papal discourse with the world.

Marking the encyclical's thirtieth anniversary in 1993, *National Catholic Reporter* columnist Thomas E. Blackburn lamented that conflicts in Bosnia underlined the challenges of minority populations and that disparities in wealth among nations were only haltingly addressed by the North American Free Trade Agreement. "Good Pope John believed that everybody could understand the principles when they were laid out for them. So he addressed people of goodwill, rather than the enforcement arms of the church. He preached, and he got a response from the wide audience. The text from which he spoke holds up better than either of the two great political-economic systems to which he preached."

In 2003, on the fortieth anniversary of the encyclical, war was being waged in Iraq. *New York Times* columnist Peter Steinfels used the occasion to question "the total pattern of American policy." With that in mind, it is fair to ask, Was *Pacem in terris* praised but ignored by warring nations? Was it overly optimistic, as theologians Reinhold Niebuhr and Paul Tillich asserted? What are the fruits of the various *Pacem in terris* peace institutes and retreat centers that have sprung up in New York, Delaware, and Minnesota?

Perhaps the most enduring legacy of the epochal encyclical is that it changed the conversation. It took the pope off his throne—at least for a significant moment in time. He engaged the world in gentle, impassioned, fatherly dialogue, understandable to superpowers and to peasants alike. His cry for true and sustainable peace based on the gospel precepts of charity, justice, and truth, with duties flowing from rights, is seen today as a given in most civilized circles. But its application in a real and change- able and too often violent world remains as elusive as ever.

CHAPTER FOURTEEN

Finis (May–June 1963)

Aside from the fruits of Vatican II, *Pacem in terris* was perhaps Pope John XXIII's greatest contribution to the world. After it was published, he concentrated on the business of living in the shadow of "Sister Death," as he always called mortality. A sculptor who had done several busts of John was disheartened when he saw the pope on Holy Saturday 1963: "Most of [his face] had fallen, except the big hooked nose and the immense ears which were left to ride above all else like an alarming sentinel, gaunt towers of a crumbling castle."

Even the pope's legendary sense of humor was tempered with an acceptance of reality. Shortly after Easter, he received a visit from the parish priest of Sotto il Monte, Father Pietro Bosio, who sought a blessing for a seminary that was to be built near John's ancestral home. Bosio told the pope that a good many villagers wanted to come and see him, to which John replied, "Well, tell them to come quickly. Are they waiting until I am dead?" Later, he approved a model of the proposed seminary, but then added, "If you hurry up and build it, maybe I will come personally and dedicate it."

In his journal, he wrote of "unbroken pain that makes me seriously wonder about my chances."

As he went about his daily business throughout the remainder of April, John balanced life with imminent death. A visit to an orchestra concert on Saturday, April 20, turned what should have been an enjoyable moment into "seventy-five minutes of pain." When Bishop Pericle Felici, secretary of the council during the first season, visited him, John turned emotional. As Felici recounted:

> He wanted to make me a gift of his book on Radini-Tedeschi, and read out extracts from it. He said: "My Bishop Radini-Tedeschi would have made a very good secretary of the ecumenical council." Then with tears in his eyes he evoked the death of "My Bishop." It was an anticipated account of his own death little more than a month later. . . . When I looked at the dedication, I was moved to see that he had written: *Ubi patientia, ibi laetitia.* (Where there is patience, there is joy.)

While John was quietly going about the business of dying, the earthly realm kept intruding. On April 30, following a national election, the Communists picked up numerous offices, winning about one million more votes than they won in the previous election, five years earlier. John's critics attributed their gains to his influence—a Milan newspaper retitled his encyclical *Falcem in terris* (The Sickle on Earth). That night, John suffered a severe internal hemorrhage that required three blood transfusions.

Still, incredibly, the eighty-one-year-old pontiff kept to his schedule; however, saying Mass the next morning, he lost his way in the liturgy, for a period forgetting the words that had come so naturally to him for almost sixty years.

He then met with James McCone, director of the Central Intelligence Agency, who warned about getting too cozy with the Russians. But John wouldn't have it. "I'm not going [to be] put off my stroke by the unseemly fuss that some people try to impress churchmen with," he wrote in his journal. "I bless all peoples, and withhold my confidence from none."

On Friday, May 10, Antonio Segni, president of Italy, presented John with the Balzan Prize in the Sala Regia, the Vatican's private throne room, rather than in Saint Peter's, because John didn't think it was appropriate for a pope to receive an award in a Church. But he was pleased with the prize. "The desire for a just peace," he remarked, "has entered the hearts and minds of all without distinction." Later, he received the scroll commemorating the prize in the basilica, using the opportunity to pray for peace. There, he noted: "Peace is the greatest treasure of life in a community, the most luminous point in the history of humanity and Christianity, the object of the trustful expectations of the Church and the people."

After the reception, when he was carried out in his sedan chair, the *sedia gestatoria,* he put his head in his hands. The ceremony— his last major address in Saint Peter's—had exhausted him. That afternoon, he was given a clipping from the *Baltimore Sun,* which cited "a Vatican staffer well placed to know" who had said, "Recently the pope has been waking up during the night and asking for sedatives."

As it happens, the Roman pontiff had not asked for sedatives, despite his great pain. After reading the article, the pope merely shook his head and told Capovilla, "File it away. Every detail helps to document situations and the movement of minds." Yet the pain continued and worsened. The next day, at a reception at the Quirinale Palace, where Segni honored John and American historian Samuel Eliot Morrison for their Balzan Prizes, John barely made it through the ceremony.

Back at home, while watching coverage of the event on television, he told Loris Capovilla, "A few hours ago I was being feted and complimented, and now I'm here alone with my pain. But that's all right. The first duty of a pope is to pray and suffer."

The reception was his last public appearance. John continued to suffer, and he continued to work. By mid-May, the pain had become so great, he told Capovilla, that he felt "like Saint Lawrence on the grid-iron." He was now sleepless much of the night. He celebrated his last Mass on May 17. Three days later, he met with Cardinal Stefan Wyszynski of Poland, who suggested the pope receive him in the papal bedroom. "We haven't come to that yet," John said. Afterward Wyszynski told him he would see him in September, at the second session of the council.

"In September you will either find me here, or another," John responded, smiling. "You know, in one month they can do it all—the funeral of one pope, the election of another." That afternoon he experienced several small hemorrhages and received two blood transfusions. The next day, after another sleepless night, he took Holy Communion at six in the morning and expressed to those present a sentiment that was increasingly common to him in his last days: "I'm ready to go. I've said all my breviary and the whole Rosary. I've prayed for the children, for the sick, for the sinners. . . . Will things be done differently when I'm gone? That's none of my business."

And yet he did not die.

One of his doctors, Antonio Gasbarrini, said that he had "a constitution of iron." On May 22, he fainted in his quarters while dressing. He was scheduled to visit the Benedictine Abbey at Monte Casino the following day, but his handlers told him he was far too weak to travel, despite his protestations. Capovilla told him that there would be little anyone could do if he hemorrhaged there, to which the pope replied with some excitement: "I'd go to bed. I'd go to a cell in the Abbey. Think of it: to die at Monte Casino, the cradle of monasticism!"

Although the Vatican continued to send out news releases about his "gastric troubles," most people understood the pope was dying. The world prayed for him.

On Monday, May 27, his nephew, Monsignor Giovanni Battista Roncalli, came to visit so that, through him, John could tell all of his extended family how much he loved them. Battista was also entrusted with telling the family about his last will and testament, in which he writes, "Born poor, but of humble and respected folk, I am particularly happy to die poor, having distributed, according to the various needs and circumstances of my simple and modest life in service of the poor and of the holy Church which has nurtured me, whatever came into my hands." Therefore, to his family, he "can leave only a great and special blessing."

Despite what was obviously intended to be a last conversation with his nephew, John told him before he left, "So you come here today and find me in my bed. . . . But let's hope I'll get over it soon and be able to get back to work on the council."

The following day, May 28, *L'Osservatore Romano* finally printed the truth about the pope's condition, admitting that he had cancer rather than "a distress of the stomach." The close observer knew by that time that the Holy Father had only days remaining in his life.

On the morning of Thursday, May 30, John was seized by sharp and violent abdominal pains, most likely the result of a perforated intestine caused by the tumor. For the first time, John was given sedatives. Peritonitis was sure to set in. The next day, John's two primary physicians, Dr. Antonio Gasbarrini and Dr. Piero Mazzoni, confirmed there was nothing they could do. The dutiful, distraught Loris Capovilla took it upon himself to tell John.

Approaching the pope's bedside, trying not to cry, he said, "Holy Father, I'm keeping my promise: I have to do for you what you did for Monsignor Radini at the end of his life. The time has come. The Lord calls you."

"It would be good to have the doctors' verdict," John answered calmly.

"Their verdict," Capovilla replied, "is that it's the end. The tumor has done its work."

"So, as with Monsignor Radini . . . there will be an operation?"

"It's too late," Capovilla answered. "The cancer has at last overcome your long resistance."

And at this point, the devoted secretary broke down weeping, his face buried in the bedspread. But John simply said, "Help me die as a bishop or a pope would."

To die as a bishop or a pope would, with due ceremony and solemnity.

History, "the great teacher of life," would provide John with a way toward his death. The pontiff told Capovilla to gather everyone around him, and soon the room was filled with his doctors and nurses; his housekeepers, nuns from Bergamo; and clergy, including Bishop Angelo Dell'Acqua and Cardinal Antonio Samorè, representatives of the Secretariat of State). John was able to sit up in bed and receive the Eucharist from Monsignor Alfredo Cavagana. As part of the last rites, Bishop Peter Canisius van Lierde, the papal sacristan, started to anoint John with oil, but the pope stopped him so that he could speak:

> The secret of my ministry is that crucifix you see opposite my bed. It's there so that I can see it in my first waking moments and before going to sleep. It's there, also, so that I can talk to it during the long evening hours. Look at it, see it as I see it. Those open arms have been the program of my pontificate: they say that Christ died for all, for all. No one is excluded from his love, from his forgiveness. . . .

Moving most of those present to tears, John thanked God for being born into a "Christian family, modest and poor." He listed those who had influenced his early life, from his old parish priest, Father Francesco Rebuzzini, to Bishop Giacomo Radini-Tedeschi and Cardinal Andrea Carlo Ferrari. One can imagine scenes from his life flooding back to John as he lay in bed—the hilly village of Sotto il Monte, swept by the *tramontano;* his early experiences in the tiny school; life with his large, boisterous family, especially his mother, Marianna; his early days in Bergamo; and finally his papacy in Rome.

So many of the people he met and loved along the way were inspirations, he told the somber group around his bedside. They "helped me and loved me. I had lots of encouragement." He offered an apology to anyone he might have offended and then told the assembled, "My time on earth is drawing to a close. But Christ lives on and the Church continues his work. Souls, souls."

Van Lierde then began to anoint the pope with the final sacrament, touching each of the five senses, but he was so emotional that he forgot the proper order. John calmly reminded him—eyes, ears, nose, mouth, hands, and feet. After this, John was able to speak personally with everyone present—about twenty people. By this time it was 4:30 in the afternoon, and now members of the Curia came through, with John bidding each one good-bye. He told one, "In my opinion, my successor will be Montini. The votes of the sacred college will converge on him."

By early that evening, the audiences were over and John was left alone with Capovilla. He spoke to him with great affection and sincerity, as if they were comrades who had survived a war together. "We've worked together and served the Church without stopping to pick up and throw aside the stones that have sometimes blocked our path. You've put up with my defects and I've put up with yours. We'll always be friends. . . . I'll protect you from heaven. . . . When this is over get some rest and go and see your mother."

The pain grew sharper, and John was put under sedation. While he slept, his brothers Zaverio, Alfredo, and Giuseppe, accompanied by their sister Assunta, came in and knelt by the pope's bed to pray. They were joined by the pope's nephews, Monsignor Battista Roncalli and Zaverio Roncalli. Midnight struck. It was June 1. A huge crowd began gathering in the square outside the window. Father Capovilla said Mass for those in the room. As Battista later wrote, the family was very aware that it was not just their Angelo dying, but the pope of the Universal Church:

> The first thing they were told was that they must not weep.
> If it turned out they could not hold back their tears, they
> were to leave. These four poor old souls were trembling and
> still upset from their airplane trip, which was the first for
> any of them. There was only a dim light in the room. We

are all standing back because we were given the impression that the pope was having great difficulty breathing, that he needed air, and that if we stood too close we would deprive him of it.

At about three o'clock that morning, the pope regained consciousness and seemed to rally. He was able to sit up, take a cup of coffee, and speak conversationally to his family. "I'm still here when yesterday I thought I was gone," he told them. "I don't know what's going to happen. I could get better. We're made to live. . . . If I get worse, what a disappointment for you." He dozed off, but woke up again within an hour, asking his brothers and sister, "Do you remember how I never thought of anything else in life but being a priest? I embrace you and bless you."

John saw a few more people during the course of the day as a crowd continued to gather on Saint Peter's Square and the world waited in vigil. During the night of June 2–3, John fell back into unconsciousness. In Milan, 20,000 young people flocked to the cathedral to pray. Cardinal Montini met with them. People needed to "gather up his inheritance and his final message of peace," the cardinal said. "Perhaps never before in our time has a human word—the word of a master, a leader, a prophet, a pope—rung out so loudly and won such affection throughout the whole world."

On Sunday, June 2, John awoke before noon with a fever of 104 degrees. He told Dr. Mazzoni, "I am suffering with love, but with pain, too, so much pain." Trying to find a parting gift for Mazzoni, he finally handed him his fountain pen. "Take it," he told the doctor. "It's nearly new." Seeing his nephew Zaverio standing at the foot of the bed, he told him gently to move: "Out of the way, you're hiding the crucifix from me."

In his bedroom, during the final hours, many crowded around the dying pontiff:

Tisserant, Masella, and Cicoganani; John's three brothers and his sister; three of his nephews, Giovanni Battista, Zaverio, and Flavio; four of his nieces, two of whom, Angela and Anna, were nuns; his two faithful valets, Guido and Giampalo Gussi; the three doctors, Cavagna and van Lierde, the sacristan; Federico Belotti, his principal nurse; the five nuns of the household; the devoted Capovilla and, in the shadows, Montini.

By nine that night, the pain had gotten a good deal worse, and John was given sedatives again. The prayers for the dying were started, but the pope continued to breathe, though not without the aid of oxygen. On Monday morning at 3:00 John awoke and said twice, very firmly, the words of Saint Peter: "Lord, you know that I love you." These were his last clear words. He slept, tossing and turning, for the rest of the day, but by early evening he had fallen into deep unconsciousness.

It was a warm spring evening outside on Saint Peter's Square. Cardinal Luigi Taglia, vicar of Rome, said Mass for John in front of a crowd of thousands. The liturgy could be clearly heard inside John's bedroom, where Capovilla, his family, and the doctors and the household staff continued to gather. They said the prayers for the dying, which rose and intermingled with the words of the sacred rite. Just before eight o'clock, Cardinal Taglia uttered the traditional sentence that ended the service: "*Ite, missa est.*" (Go, the Mass is ended.) Just at that moment, John took his last breath.

Those in the room with John knelt and prayed and then sang hymns—the Te Deum and the Magnificat. As tradition dictated—and tradition was something John loved—the dead pope's brow was tapped firmly, to make sure he was really dead. Then the shutters to the "window of the Angelus"—the window where popes appear to pray the Angelus—were opened. Light flooded out and those on the square knew that John was dead. To the world at large, the Vatican press office issued a statement: "He suffers no more."

Although the whole world had followed John through his illness and death, his dying had remained a relatively private affair. This was in part because, on September 2, 1962, he had signed a personal statute entitled *Summi Pontificis electio,* which decried certain "facts and customs recently introduced." This was a reference to photographs taken after the death of Pope Pius XII, as he lay in his bed with medical devices attached to his body. John had decreed that no photographs or film or voice recordings were to be made in his apartment after his death, except for one official photograph, for historical purposes.

But John's funeral and burial were public. On June 4, his body was carried through Saint Peter's Square and placed in the basilica, where for two days thousands filed past his bier in an unending line. On June 6, he was interred in the crypt below Saint Peter's. Those who loved him—and they numbered in the millions—were happy that he was out of pain and had gone for his audience with the Lord. He had not been a man who wanted to die, but he was ready to, as he told Dr. Mazzoni near the end: "Don't look so worried. My bags are packed and I'm ready to go." He had been pope for just under five years—up to that point, the shortest papal reign since Pius VIII—but he had accomplished an extraordinary amount.

Whatever the ultimate effect of Vatican II—and its virtues and demerits are still being debated today—John himself was an extraordinary man and pope. Many thought him a saint. At the second session of the council, his close friend, Belgian Cardinal Leo Suenens, thought that John should be canonized right then and there, by acclamation, as had been done in the early years of the Church. He and several bishops circulated a petition that read, in part:

> From Pope John the world has learned that it is not so alienated from the Church after all, nor is the Church from the world. Maybe now the world expects us to declare that we do not consider Pope John a dreamer, or as one who had rashly overturned everything that, with long and patient effort, we will have to put back in order . . . but on the contrary, that we see him as a true Christian, indeed a saintly one, a man filled with love for the world and for all mankind.

However, the Curia intervened. It would not do to make John a saint by acclamation—this would make his predecessor, Pius XII, look bad by comparison. The Vatican Congregation for the Cause of Saints also firmly disagreed, since the petition would in effect take power out of their hands—*they* were the ones who picked saints, after all. And so the saint-by-acclamation movement died a relatively quick death. John was beatified, however, by Pope John Paul II in 2000 (John Paul himself, on a considerably faster track, was declared Blessed in 2011); it remains to be seen whether the Church will grant him sainthood. Most of those who knew him well thought that he was a saint, in the sense of his being a holy man who worked tirelessly to improve the lives of others and who believed firmly in salvation through Jesus Christ.

Each candidate for sainthood is assigned a postulator— someone who conducts investigations into the life of the would-be saint in order to ascertain whether he or she is worthy. Father Luca M. De Rosa, an Italian Franciscan, is postulator for John's cause. "Believers but also nonbelievers" admire him, says De Rosa. "They continue to admire Angelo Roncalli's goodness and mild- ness, his zeal for truth, peace and understanding between peoples, his anxiety to be God's messenger and a servant of humanity. . . . His life story makes people want to live better. . . ."

It was a benediction that John would have loved.

At the consistory of December 15, 1958, the first cardinal John named was Giovanni Battista Montini, the archbishop of Milan, former influential curial official under Pius XII, and a personal friend of John's for more than thirty years.

Then, on September 5, 1962, a month before opening the Vati- can Council, he published the motu proprio, *Summi Pontificis electio,* which would serve as the rules for the next conclave. The pontiff made several minor changes to Pius XII's constitution, under which he had been elected, including elimination of the provision that two-thirds *plus one* votes are required for the ca- nonical election of the pope. Also, each cardinal elector was to be accompanied by only one conclavist (assistant), and all would be clerics. There would be no valets or other laity housed within the conclave itself, other than the staff who would provide the medi- cal, nutritional, and security support needed.

In the end, the committee charged with organizing the con- clave approved 80 electors, 82 conclavists, and 119 staff members.

Fifty-four votes were required for election. It turned into a classic battle of the curial conservative-traditionalists (who were opposed to the liberalizing movement of the council) and the liberal or progressive majority (though it would be a struggle to garner two-thirds behind one candidate, as always, because conservatives controlled more than one-third of the total votes).

Montini of Milan, age sixty-five, was the clear favorite going into the balloting. After the first ballot, he led with a plurality of twenty-eight votes. On Thursday, June 20, everyone knew that he was John's choice as a successor, but Church law and tradition forbade the pope from dictating his successor.

In all, it took less than the month John had predicted to elect a new pope. On the following day, Friday, June 21, Montini was elected on the sixth ballot. Just over three weeks after John's burial, Cardinal Montini stepped out onto the balcony as the newly chosen Pope Paul VI, to the thundering applause of the huge crowd gathered in Saint Peter's Square. He was, in effect, John's handpicked successor, and despite rumors that he had been "very pale, almost speechless" when he found out he had achieved the necessary votes in the conclave, he started out vigorously and was firm in his support of John's council, setting the opening date of the second session for September 29, 1963.

"We wish our thoughts also to go to those of our brethren who are not Catholics," he told a group of American Catholic pilgrims on June 25. "On them and their dear ones, we evoke the abundance of heavenly grace." Even the name Montini chose—Paul, the apostle to the gentiles—was evocative of the ecumenical spirit that had so imbued John and his council, which was to finally end after its fourth session in 1965.

Chapter Fifteen

The Good Pope
and His Great Council

O
n Sunday, November 27, 2011, the Catholic Church in the United States implemented a new translation of the Roman Missal, the "textbook" of Roman Catholic worship that contains all the prayers of the liturgy said by the priest and the people. The reason the U.S. Conference of Catholic Bishops revised the English translation was to bring the language of the Mass closer to the source from which it had come: the post–Counter-Reformation Latin of the one, holy, Catholic, and apostolic Church of which they are a vitally important regional body. In essence, it was a conservative move "back" to a more "faithful" translation and interpretation of the traditional texts of the Mass.

Arguably, however, and despite the direction of the change, it was also a direct result and manifestation of the movement and spirit of Vatican II, which radically altered the liturgy for all Catholics.

This recent change started American Catholics talking about the liturgy—and their participation in worship—again, in a way not experienced in North America since the 1960s and 1970s at the height of the postconcilliar implementation of the reforms that the Good Pope had instigated. So, in effect, the hand of Pope John XXIII was being felt in a very substantial way some fifty years out and by new generations of U.S. Catholics who were unaware as to the revolution that had swept through their Church in that faraway time.

Half a century earlier, it was not at all clear that the Second Ecumenical Council of the Vatican would ever really happen. It was seen as a historical gamble, and bets were still being placed on the table. Nonetheless, John had forged ahead and required of his brother bishops and his curial mandarins that they come along with him. He *moved* the Church in ways still felt today, five decades later—and if he had not, it is impossible to know what the state of that Church and our world might be, nor how a billion souls would be nourished with a Word that claims eternal potency and absolute truth.

In theological terms, John responded to the prompting of the Holy Spirit by convoking the Second Ecumenical Council of the Vatican; he was clearly inspired, in the purest sense, by the third person of the Holy Trinity. Those who invested much hope in the promise of the council, Catholics and non-Catholics alike, were perhaps inevitably disappointed at the outcomes as the months and years rolled along. But so too were those who were skeptical of the council's purposes: They objected to the lengths to which the synod stretched its mandate for reform, carping that it had overextended that mandate.

In other words, not many were completely satisfied by the council's legislation, and conversely, many were disturbed by the wrenching changes wrought by the fathers—particularly in liturgical matters—even when such changes had received overwhelming votes on the floor of the assembly.

So, is this how the Holy Spirit is supposed to work in the lives of men and women, in the life of the Church? John and his adherents would claim that it is, prima facie.

In a role unique to the modern papacy, though he never led a parish church nor was he renowned for his academic prowess, John melded the temperament of a true pastor with the intellectual activity of a natural theologian. He was not a slouch at either. He came to the Throne of Peter via a career in diplomacy, just as his predecessor, Pius XII, and successor, Paul VI, did. And that is what he was, too, in spades. Who else could have kept the ship of the council afloat through the stormy first session, let alone have launched it and assuaged (at least some of the time) the egos and hugely disparate priorities of the experienced officers on his deck?

Somehow, perhaps with elements of the miraculous, certainly with personal magnetism and powers of persuasion, the peasant-born pope leveraged the power and pomp of his ancient office to see that things got done his way, more or less. As pastor, he knew that he must lead by influence and be seen as a true servant of the Servants of God. As theologian, he knew that he possessed the authority, albeit temporarily and by the spiritual accident of his election, and that he must persuade with the Word and with his own words.

In any number of instances, he let the Roman Curia have its way. To Cardinal Richard Cushing, in this regard, he confessed, *"Sono nel sacco qui."* (I'm in a bag here.) But the right hand worked constantly, even as the left hand seemed to let go of the strings of papal power:

> When the council fathers arrived in Rome, they began get-
> ting discreet telephone calls from Monsignor Loris Capo-
> villa, the pope's private secretary, subtly disassociating the
> pope from the Curia. The progressives among the bishops
> correctly deduced that John wanted a wholesale reform, but
> they did not at first realize their own strength. Gradually,
> encouraged by the knowledge that the world was watching,
> they became emboldened.

He emboldened others through his own adherence to discipline and his willingness to tolerate debate and speculation in
the cause of finding the ultimate, if not always immediately apparent, truth.

The first session of the council is remembered for the battles
fought over three important schemata, the proposals of the preparatory commissions that had caused such a kerfuffle just days
after the session finally opened.

First, by a vote of 1,922 to 11, the council approved the liturgical
reforms that enable the world's bishops, in their various regional
conferences, to decide the language of the Mass to be used by
their people. This move, later codified in the Constitution on the
Sacred Liturgy, *Sacrosanctum concillium,* the first major piece of
Vatican II legislation, was one of decided decentralization, if not
democratization, that set the tone for the subsequent three years
of debate and decision making by the fathers.

Another critical debate involved the sources of revelation, the
controversial schema presented by the redoubtable Cardinal Ottaviani, who was (in simplified language) the "anti-John" of the

Second Vatican Council. While ever protective of the prerogatives of the bishops and the pope, who is also a bishop, in their unchallenged function as the Magisterium, or ultimate teaching authority, Ottaviani insisted that the two sources known as tradition, pre- and post-scriptural teachings handed down orally, and scripture, the written words revealed to the authors of the Bible, be recognized as separate and equal—contrary to Protestant Christian reliance on *sola Scriptura,* "scripture alone." Shelving the conservative Ottaviani's proposed document, the council moved away from the Counter-Reformation tendencies of the Church since the sixteenth-century Council of Trent and toward John's vision of a modern institution engaged with the world outside Catholicism's "boundaries." He had lived the concept of ecumenism and brought it into legitimacy as never before.

Finally, the fathers debated the nature of the Church during the first session and throughout the balance of Vatican II. Again in response to Ottaviani, the progressive council members wanted this issue—a frank discussion of what the Church does and who she is, which had been begun, then suspended, at the abortive First Vatican Council—on the table. The so-called liberals wished to purge the Church of the stains of triumphalism, clericism, monarchism, and militarism that had caused the institution to stand apart, oftentimes impotently, from the world around it and even from its own faithful. This debate opened the way for serious reconsideration by the fathers of the Catholic Church's position on religious freedom, of Church-state relations in countries around the world, and of opening the eyes and the arms of the Church to her own laity, reaffirming the "universal priesthood" of the faithful. For many Catholics, especially in North America, this was the major accomplishment of the council. More than ever before in the history of the Church, the laity were invited to take a more active, visible role in the life of the Church and its ancient liturgy.

Each legislative act in itself was a tall order, broaching sensitive subjects that had not been spoken of in countless centuries, if ever before. John stood by during the debates, observing with some pleasure and perhaps some sadness the ebb and flow of theological development that proved his Church was neither moribund nor monolithic.

The debates and the direction of his council also proved John to be an intuitive leader open to the inspiration of the Holy Spirit in the moment of an event or at the birth of an idea. In his manner, which reflected the temper of his soul, this pontiff of 1,680 days inspired others to dream, to talk, to act, to be Catholic in whatever state of life they might find themselves.

When *TIME* magazine named John its "Man of the Year" in 1962, the magazine noted:

Though Pope John had proved a happy surprise to both the Catholic Church and the world, his life is full of signposts that clearly mark his life and growth. He is an intuitive being who can pierce to the heart of the matter without taking the circuitous route of deeper and more discursive minds. The rhythmic natural influences of his first years on the farm at Sotto il Monte formed him for all time. A few weeks ago, asked by some bishops what he wanted to do after the council, John replied: "Spend a day tilling the fields with my brothers." Neither an intellectual nor a highly trained theologian, he does not think in concepts but in terms of fundamental human experiences. In a varied and unusual career, he has absorbed and synthesized these experiences to an extraordinary degree.

He denied some of the salient and varied aspects of his character that were apparent to others—out of a deep humility that seemed to spring from his DNA. He really wanted to be remembered, first and only, as "the good shepherd defending truth and goodness." And despite his roly-poly figure, he walked forth to many places outside the confines of the Vatican, famously to jails, orphanages, government offices, schools, and, of course, churches some 139 times in four and a half years.

Uniquely, John was a pastor to those outside the confines of the Church, as well. Throughout his life and ministry he sought out non-Catholics and non-Christians, an unprecedented act for most popes but a signature trait of his papacy. Ever since, his successors have carried on this tradition with relative vigor over the half-century since the Good Pope's passing. While in Turkey, John helped rescue and provide for Jews escaping from Nazi Germany. In France after the war, he recoiled in horror when he saw films of Jewish bodies piled high at Buchenwald and Auschwitz, saying, "How can this be? The mystical body of Christ!" Memorably, when a group of Jews visited him after he became pope, John walked up to them and simply repeated the biblical greeting, "I am Joseph, your brother."

Thus, in his public actions, he lived out the gospel among Jews, Muslims, and other non-Catholics, and "anyone who does not call himself a Christian but who really is so because he does good."

The body of the late John XXIII was transferred from the crypt beneath Saint Peter's Basilica, where he presided over the Second Ecumenical Council of the Vatican. Forty thousand people attended the reinterment ceremony on June 3, 2001, thirty-eight years to the day of his death.

The only other popes so memorialized and publicly displayed under glass are Blessed Innocent XI, who died in 1689 and was beatified by Pius XII in 1956, and Saint Pius X, who died on the eve of World War I and was canonized in 1954, also by Pius XII.

When his coffin was reopened, his physical remains were remarkably uncorrupted. After less than a day of work on the corpse, those present saw the face of John XXIII. Cardinal Virgilio Noe, who was in charge of the project as the archpriest of Saint Peter's Basilica and president of the congregation responsible for the "physical plant" of the Vatican, described John's face as "intact and serene." He said witnesses, present at the opening of the coffin, were overcome with emotion. "It is a providential coincidence," he said, "a sign of divine favor and of holiness."

Some wanted to call it a miracle, proof of the Good Pope's sanctity. Dr. Gennaro Goglia, an anatomy professor who, in 1963, had injected the pope's body with a special embalming fluid, eschewed any discussion of miracles. In fact, Goglia described the procedure in this way:

> We put the bottle containing the liquid on the tripod. We made a small cut in the right wrist and inserted the needle there. I was afraid that the blood would exit through the tube or that the liquid would cause the skin to rupture. . . . At 5 A.M. on June 4 [1963], the operation ended. The liquid had reached all the capillaries, blocking the process of decomposition. We then injected some liters of the liquid into the pope's stomach, destroyed by cancer, in order to kill the bacteria there.

A layer of protective wax was applied to the face, which was nonetheless unmistakably that of Roncalli.

Prior to the Pentecost Sunday Rite, the body was vested in new papal garments of silk, his head covered with a cap bordered with ermine, a trademark look for John. A new ring was placed on his finger. His body was placed within a new bronze-and-glass coffin, then carried into Saint Peter's Square. It was reportedly the first time a pope's remains were venerated in the open air. After a solemn outdoor Mass, the coffin was then transported into the basilica, where it was permanently placed beneath the Altar of Saint Jerome in the central nave.

The miracle of John's Second Vatican Council, which became Paul's council and the Church's council in the end, is that it occurred. If John had not acted upon the inspiration of a moment— of many moments in his life, of the Holy Spirit acting upon *him*—the Catholic Church might have become calcified. The Church had survived worse, much worse, than anything that happened in the past few centuries, even the past couple of decades. But John understood that there was no time to lose in moving the Church to reform; events in the world had accelerated beyond what previous generations had experienced—and would continue to do so at a faster and faster clip for the coming fifty years.

From the beginning, as early as 1959, some members of the hierarchy and within the Church at large have challenged the reforms of the council and resisted the movement toward ecumenism, including relations with other Christian denominations and non-Christian religions. The defensive walls of the ancient fortress were breached in the early 1960s, but new walls within the structure have been built over time.

However, the purpose of John's council was not to make a radical break with the past or to put aside any doctrine. Remember that John himself—Angelo Roncalli of Sotto il Monte—was deeply traditional in matters of faith and morals and orthodox in his understanding of the Catholic faith and Church dogma. Yet he appreciated the need for the theological development and engagement in the contemporary world through the social teaching that had developed in the Church since Leo XIII. His council was to be the most important moment in the history of the Church since the Council of Trent. If nothing else, the liturgy would be reformed as one of the most enduring and outwardly visible signs of the life of the Church for all to see.

Cardinal Alfredo Ottaviani of the Congregation of the Holy Office and Archbishop Marcel-Francois Lefebvre of Tulle, France, were two opponents of John's council—and of Paul VI's agenda as the successor of the Good Pope. The former worked to reshape the direction of the council legislation every step of the way, until his retirement in 1968. The latter forced a schism under the aegis of his religious order, the Congregation of the Holy Spirit, rejecting Vatican II outright, before his death in 1991.

More subtly and steadily through the decades since the closing of the council in 1965, some bishops and cardinals throughout the world have attempted to mitigate the pace and depth of reforms in their dioceses and the curial agencies. Nevertheless, the council lives on in its teachings and will, over time, continue to be absorbed into the fabric of the Church, just as all such events have throughout the past two millennia. The most obvious and enduring reform, for example, is the liturgy of the Eucharist (the Mass), now being celebrated around the world in the local vernacular and with local cultural markings.

In his moment of history, John stood at the pinnacle of an ancient religious hierarchy during an era of secularism that held the very real potential for mass destruction of peoples across the globe. With both a prayerful humility and an iron will, he moved a massive, entrenched institution toward a more open relationship with and engagement in the world through the gospel. In his person, he represented the dignity of the human being as—in the doctrine of his Church—having been created by God in the divine image and likeness.

Unblinking, and with a smile, he sought to reform and reclaim his cherished Church. Counter to the ecclesial culture that had formed and nurtured him, John emerged as a leader with a surprising agenda that threatened that culture and the institution to which he had given his life. His was an essentially conservative mission to cure and preserve life that required radical surgery to achieve its true end. He worked, masterfully, within the structures in which he had been raised to the pinnacle of power to answer what he felt—or knew—to be the prompting of the Holy Spirit.

In his lifetime he stood against powerful forces of history within and outside Catholicism, and his legacy would be met with efforts at retrenchment—from reactionary elements among both clerical and lay segments of his Church. Nonetheless, he did not stand alone. His favored successor, Giovanni Battista Montini, Pope Paul VI, had assiduously supported John's initiative and reforming mission as a leader at the council and the logical executor of the project after John's passing. Cardinal Augustin Bea, and others, also seized the opportunities John presented to them and carried forward the spirit of *aggiornamento* into all corners and recesses of the Church, high and low, far and wide.

The people (perhaps the laity and religious communities more than many of the members of the hierarchy), in the final analysis, have adopted John's Church and adapted to it. They are still in the process of absorbing the meanings of the changes along with the enduring verities of his teachings.

The key to John XXIII and his remarkable council, ultimately, is the unity of vision for the Church and her people expressed in the documents that emerged from the historic synod. What took decades to emerge from the deliberations of the Council of Trent—sweeping reforms of liturgy, education, and discipline—took only four years at Vatican II, easily the largest and most diverse council body in the history of the Church.

As Adam was created from the dirt of Eden in the biblical narrative of Genesis, Angelo Roncalli emerged from the mountain soil of Sotto il Monte. Throughout his papacy, John remained a vital, robust, earthy character whose spirituality was simplicity itself. He loved scripture and praying his daily Office, but he equally loved and drew sustenance from the tradition of his Church. Although he traveled far into Eastern Europe and to the doorstep of Asia for his diplomatic assignments, he retained the insularity of his native village and the interiority of a village priest—while robed in the outer garments of worldliness and sophistication he put on and took off with ease. Whether in Paris salons or on the streets of Istanbul or the rural hamlets of Bulgaria, he was always comfortable in his surroundings; even in the precincts of the Vatican he found a home among the priests and workers during his short reign as Supreme Pontiff of the Universal Church.

The Johannine spirituality that emerged in the final decade of his life seemed suited for the time. In the morning of the post–World War II age and the increasing chill of the Cold War, he expressed simply and eloquently religious concepts of human dignity, compassion, and peace in the language of the Church and in words and images understood everywhere throughout the world by Catholics and non-Catholics alike.

He also articulated his vision for the Church in his overt activities. His religious standard was always conventional and orthodox, though from his earliest years as a priest and teacher in seminary and university, he exhibited a capacity for inclusivity that broadened, almost immediately upon his election as pope, to incorporate the world beyond the Church and its many religions—something for which no other previous Bishop of Rome would ever be remembered.

Other popes have since worn the same red "Shoes of the Fisherman" as John, and others will in the future. It is difficult to imagine that any will be judged by history—or in the hearts of men and women around the world—as having touched his time with as much simple grace and enormous impact as the one the world still calls the Good Pope.

ACKNOWLEDGMENTS

From the original presentation of this book as an idea to its publication, less than two years has passed. That is not a very long time at all in the scheme of things—especially in the world of book publishing. But in reality, I have lived with Pope John XXIII for nearly my entire life, and I have wanted to spend more "quality time" with him ever since I was a small child and he was the reigning pontiff of the Roman Catholic Church.

The opportunity to present my proposal to the publisher was given to me by my literary agent, Stephen Hanselman. In turn, the editorial director of HarperOne, Michael Maudlin, responded with alacrity and has skillfully shepherded the manuscript through from beginning to end, with the able assistance of Miles Doyle and Kathryn Renz.

Joseph Cummins, an experienced author and longtime colleague, helped me shape the biographical narrative with great skill and sensitivity; it's accurate to say that this book could not have been produced on time without him. Paul Kocak, my previous collaborator, provided critical research on the Good Pope's writings and teachings. Mary Beth Ciccone, a Seton Hall history scholar and expert on the World War II era, dug deeply into the sources on Angelo Roncalli's diplomatic postings to Bulgaria, Greece, Turkey, and France. Ryan Dahn, a recent Princeton graduate, helped organize research into the Catholic, Orthodox, and Jewish populations of the countries to which Archbishop Roncalli was assigned.

Librarians and scholars at Seton Hall University have been immensely important to this project, including those I must acknowledge here (without their academic or clerical titles or in any particular order): Robert Wister, John Radano, John Morley, Thomas Guarino, Anthony Lee, Alan De Lozier, James Kimble, and the presidents of the university, with whom I have been privileged to work, Robert Sheeran and Gabriel Esteban.

Monsignor Charles Murphy, the former rector of the North American College in Rome, encouraged me in the early stages of research and writing and allowed me prepublication access to his own recollections of Pope John from the days he was a seminarian in Rome during the historic pontificate.

Stephen Fichter, teacher, priest, pastor, sociologist, and adviser, gave the manuscript a careful, critical reading when it was still raw and unfit for consumption by the world. And Michael Walters, another priest friend and reader par excellence, saved me from errors of fact and tone.

From authors and scholars who have written about the life and legacy of John XXIII, I have drawn extensively, attempting to fuse previous generations of research with my own and with the perspective, now, of a half-century of history since John passed from the scene.

PART I: PRIEST AND PROTECTOR

The sources for the first chapter and the entire book consist primarily of biographies, memoirs, and contemporary news accounts of Angelo Roncalli and his Catholic Church. See the bibliographical listing, "Sources," for complete details.

One important source for the author's reflections in the first chapter was *The Modern Papacy Since 1789* by Frank J. Coppa, which is a scholarly account of popes and the papacy since Pius VI.

Throughout this book, Peter Hebblethwaite's magisterial and authoritative biography, *Pope John XXIII: Shepherd of the Modern World,* is drawn upon heavily and cited with some frequency.

The parish register entry (chapter 2) is quoted in Pope John's own *Journal of a Soul,* which is quoted throughout this book. Two excellent biographies are cited throughout the chapter: *A Man Named John* by Alden Hatch and *I Will Be Called John* by Lawrence Elliot.

Both Hebblethwaite and *Journal of a Soul* also provide narrative material on the critical early stages of young Roncalli's life, family situation, and experience in formation for the priesthood.

For an accounting of his first decade as a priest (chapter 3), the major sources include Elliot, *Journal,* Hebblethwaite, and Hatch.

The primary quotes in the chapter dealing with Roncalli's World War I experience and the years immediately after (chapter 4), are found in Hebblethwaite's biography, with additional material gleaned from Hatch and Elliot.

Archbishop Roncalli's own reminiscences are recorded in *Journal of a Soul,* as well as by various biographers and journalists. Previously cited biographies cover his diplomatic career in Eastern Europe quite fully and candidly. The crucial World War II years are surprisingly not covered as widely or deeply as one would hope, but I found two sources that provided keen insight and from which I drew material: *Vatican Diplomacy and the Jews During the Holocaust, 1939–1945* by John F. Morley and *The Holy See and the War in Europe: March 1939–August 1940* by Gerard Noel.

The diplomat's time in France (chapter 6) and the patriarch's assignment to Venice are well documented from contemporary sources such as the *New York Times* and *TIME* and *Life* magazines (the latter with consistently fascinating pictures of the man and his milieu that continue over the next decade-plus into his pontificate).

PART II: THE SOUL OF A POPE

In addition to the previously cited biographies, the account of Pope John's election as supreme pontiff and his historic first days (chapter 7) can be found in *Passing the Keys: Modern Cardinals, Conclaves, and the Election of the Next Pope* by Francis A. Burkle-Young, a masterful study of the conclave system with inside information (including ballot totals quoted in this book).

Contemporaneous accounts of Pope John XXIII's pontificate and responses to his writings (chapters 8–11) can be found in newspapers and magazines of the time, including the *New York Times* (which devoted a huge amount of space to happenings at the Vatican and John in particular) and *National Review,* both of which provide an American perspective on the Good Pope and his policies—from different, virtually opposite perspectives.

Also, the pope's longtime secretary, Monsignor Loris Capovilla, recorded his reminiscences of his beloved mentor in a memoir as well as in various interviews in intervening years.

For chapter 9 in particular, the author relied heavily on a gem of an account of the Vatican City State titled *The Vatican Empire* by Nino Lo Bello, which also includes classic anecdotes of the Good Pope, focusing on his interactions with "regular" people in the employ of the Vatican and visitors to the Holy See.

World events and preparations for the council in 1961 and 1962 (chapters 10 and 11) have been written about for decades, and the author sought accounts of papal actions and responses within and outside the Catholic Church in sources such as the *New York Times* and *TIME* magazine, which are quoted, along with Vatican documents promulgated by the pope himself.

PART III: FATHER OF THE COUNCIL

Just as the focus of John's pontificate shifted in the final several months of his life, the chapters on the Second Ecumenical Council of the Vatican (chapters 12–14) reflect that shift in relying on the literature of the council that was published at the time, immediately after, and in recent years with the approach of the fiftieth anniversary of its opening.

Reflections in the final chapter (chapter 15) similarly cite a variety of sources, including biographies such as Hebblethwaite's (drawn from ubiquitously, as before) and incisive commentaries such as those by O'Malley (recent) and Rynne (contemporaneous).

SOURCES

Alberigo, Giuseppe, and Joseph A. Komonchak, eds. *History of Vatican II*, Vols. 1 and 2. Maryknoll, NY: Orbis Books, 1997.

Blackburn, Thomas E. "After 30 years, 'Pacem in Terris' hasn't lost its glow." *National Catholic Reporter.* November 3, 1993.

Braham, Randolph L., ed. *The Vatican and the Holocaust: The Catholic Church and the Jews During the Nazi Era.* New York: Columbia Univ. Press, 2000.

Burkle-Young, Francis A. *Passing the Keys: Modern Cardinals, Conclaves, and the Election of the Next Pope.* Lanham, MD: Madison Books, 1999.

Cahill, Thomas. *Pope John XXIII.* New York: Viking, 2002.

Capovilla, Loris. *The Heart and Mind of John XXIII: His Secretary's Intimate Recollections.* New York: Hawthorn Books, 1964.

Catechism of the Catholic Church. New York: Doubleday, 1995.

Coppa, Frank J. *The Modern Papacy Since 1789.* London and New York: Addison-Wesley Longman, 1998.

Elliot, Lawrence. *I Will Be Called John: A Biography of Pope John XXIII.* New York: Reader's Digest Press/E. P. Dutton, 1973.

Flannery, Austin, O.P., gen. ed. *Vatican Council II: The Conciliar and Post Conciliar Documents.* Northport, NY: Costello, 1975.

Giovanneti, Alberto. *We Have a Pope: A Portrait of His Holiness John XXIII.* Westminster, MD: Newman Press, 1959.

Graham, Robert A., S.J. *Vatican Diplomacy: A Study of Church and State on the International Plane.* Princeton, NJ: Princeton Univ. Press, 1959.

Gratsch, Edward J. *The Holy See and the United Nations, 1945–1995.* New York: Vantage Press, 1997.

Hatch, Alden. *A Man Named John: The Life of Pope John XXIII.* New York: Hawthorn Books, 1963.

Hebblethwaite, Peter. *Pope John XXIII: Shepherd of the Modern World.* New York: Doubleday, 1985.

John XXIII. *Journal of a Soul.* New York: McGraw-Hill, 1964.

Johnson, Paul. *Pope John XXIII.* Boston: Little, Brown, 1974.

Kaiser, Robert Blair. *Pope, Council and World: The Story of Vatican II.* New York: Macmillan, 1963.

Lo Bello, Nino. *The Vatican Empire.* New York: Simon and Schuster, 1970.

McBrien, Richard P. *Lives of the Popes: The Pontiffs from Saint Peter to John Paul II.* San Francisco: HarperSanFrancisco, 1997.

Morley, John F. *Vatican Diplomacy and the Jews During the Holocaust, 1939–1945.* New York: Ktav, 1980.

New York Times. Daily issues throughout Pope John XXIII's pontificate, 1958–63.

Noel, Gerard, English-language ed. *The Holy See and the War in Europe: March 1939–August 1940.* Washington, DC: Corpus Books, 1965.

O'Grady, Desmond. "Almost a Saint: Pope John XXIII." *The American Catholic.* November 1996.

O'Malley, John W. *What Happened at Vatican II.* Cambridge, MA: Harvard Univ. Press, 2008.

Riccards, Michael. *Vicars of Christ: Popes, Power, and Politics in the Modern World.* New York: Crossroad, 1997.

Ridley, F. A. *The Papacy and Fascism.* London: Martin Secker Warburg, 1937.

Rynne, Xavier. "Letter from Vatican City." *The New Yorker.* Various issues throughout the council years, 1962–66.

_____. *Vatican Council II.* New York: Orbis Books, 1999 (reprint edition).

Steinfels, Peter. " 'Pacem in Terris,' and debate on it, echo anew." *New York Times.* February 1, 2003.

TIME. Weekly issues throughout Pope John XXIII's pontificate, 1958–63.

Tobin, Greg. *Selecting the Pope: Uncovering the Mysteries of Papal Elections.* New York: Sterling, 2003.

Wynn, Wilton. *Keepers of the Keys: John XXIII, Paul VI, and John Paul II—Three Who Changed the Church.* New York: Random House, 1988.

Index